石油科技英语系列教材

丛书主编 ◎ 江淑娟 吴松林

Oil and Gas Surface Operations
油气地面处理

李滨 高苏彤 ◎ 编

石油工业出版社

内 容 提 要

油井采出的是油、气、水及各种杂质的混合流体,必须经过分离和加工处理,才能符合管输及用户需求。本书介绍了油气是如何从各油井流入处理中心的,以及处理中心的各种油气分离设备和分离技术,具体包括原油处理系统和采出水处理系统、油气两相分离器和油气水三相分离器、天然气分离和加工技术。

本书是石油工程专业学生的专业英语教程,也可供从事油气生产和加工工程的技术人员学习参考。

图书在版编目(CIP)数据

油气地面处理/李滨,高苏彤编.

北京:石油工业出版社,2014.8

(石油科技英语系列教程)

ISBN 978 – 7 – 5183 – 0214 – 7

Ⅰ.油…

Ⅱ.①李…②高…

Ⅲ.油气处理 – 英语 – 教材

Ⅳ.H31

中国版本图书馆 CIP 数据核字(2014)第 133424 号

出版发行:石油工业出版社

(北京安定门外安华里 2 区 1 号   100011)

网　　址:http://pip.cnpc.com.cn

编辑部:(010)64523574　发行部:(010)64523620

经　　销:全国新华书店

印　　刷:北京中石油彩色印刷有限责任公司

2014 年 8 月第 1 版　2014 年 8 月第 1 次印刷

787×1092 毫米　开本:1/16　印张:21.25

字数:389 千字

定价:54.00 元

前　言

　　全球石油资源分布、生产及消费三者间存在着严重的地区失衡,中东和亚太是失衡最严重的地区,中东地区严重供过于求,亚太地区严重供不应求。因此,能源行业出现了全球化发展趋势,能源国际间的交流与合作日益密切。为保证中国能源安全,中国石油石化行业国际化和本土化发展势在必行。中国油气企业正在积极进行海外业务拓展,了解资源地区的文化背景、经济发展状况、能源开发政策以及掌握其石油地质结构、油气成藏条件、开发和炼制技术等将有利于对资源地区的油气开发和炼制,更有力地支持中国经济的快速发展。

　　自 1993 年起,针对石油院校和石油职工专业英语教材严重匮乏的现状,本丛书主编陆续在黑龙江人民出版社、黑龙江科技出版社、石油工业出版社等出版了系列专业石油英语教科书,积累了一定的编写经验、培训经验和图书项目导向经验。20 年过去了,石油行业也发生了巨大的变化,新油气资源不断发现,开采与炼制等技术不断更新,海外合作区域也不断拓宽。为了适应新形势,我们经过不懈努力通过了中国石油天然气集团公司图书出版立项,开始编写一套更大规模的《石油科技英语系列教程》,既包括石油上、中、下游生产技术的部分,也包括世界主要石油资源国的经济、贸易和文化等,目的是为读者奠定走向世界石油领域的语言基础。

　　我们深感责任重大,从中国石油大学、东北大学、东北石油大学、西安石油大学及各油田石油地质研究院、设计院等单位聘请有关专家学者,确定编写体例,搜集资料。在选材上注重内容的系统性,争取覆盖本领域主要内容;语言方面,注意遴选突出科技英语语言特点的语段和篇章,并对语言使用方法进行详尽解释,以英语基础知识和基本技能的培养为主。为降低学习难度,为每篇课文还配写了汉语译文,以提高学生的石油科技英语阅读、翻译及写作能力。

　　总之,《石油科技英语系列教程》丛书在中国石油走出去之际问世,是恰逢其时,相信该系列教程对中国石油国际化发展将起到重要的推动作用。

但是这套图书由于所涉及的专业范围非常广泛,可资借鉴的新文献资料非常稀缺,作者时间和水平有限,难免出现很多不尽如人意的地方,敬请专家和读者批评,以利于将来推出更加成熟的增强版系列,为中国石油石化系统广大师生和在职员工提高英语应用能力做出我们更大的贡献。

《油气地面处理》分册介绍了油气是如何从各油井流入处理中心的,以及处理中心的各种油气分离设备和分离技术,具体包括原油处理系统和采出水处理系统、油气两相分离器和油气水三相分离器、天然气分离和加工技术。

《油气地面处理》分册第一章至第三章由中国石油大学(北京)的李滨编写,第四章至第六章由西安石油大学的高苏彤编写。李春蕾、张宁、秦亮和黄晓艺帮助做了大量的资料查找和整理工作,在此表示感谢。

丛书主编:江淑娟　吴松林
2014 年 3 月

Contents

Chapter 1　Petroleum Production and Separation

1.1　Introduction to Oil and Gas Surface Operations

🔲 Guidance to Reading

More often, the well gives a combination of gas, oil and water and various contaminants which must be separated and processed. The production facility performs processing of production fluids in order to separate out key components for the market and dispose of waste materials to be environmentally acceptable. The production separators come in many forms and designs. In addition, the facility must provide for well testing and measurement so that gas, oil, and water production can be properly allocated to each well. Gas sales contracts require most of the water vapor to be removed from the gas to prevent hydrates and increase line capacity.

🔲 Text

Typical production fluids from oil wells are a mixture of oil, gas and produced water. The job of a production **facility** is to separate the well stream into three components, typically called "phases" (oil, gas, and water), and process these phases into some marketable product(s) or dispose of them in an environmentally acceptable manner. In mechanical devices called "separators" gas is **flashed** from the liquids and "free water" is separated from the oil. These steps remove enough light hydrocarbons to produce a stable crude oil with the **volatility** (vapor pressure) to meet sales criteria. Separators can be either horizontal or vertical in configuration.

The gas that is separated must be **compressed** and treated for sales. Compression is typically done by engine – driven **reciprocating compressors**. In large facilities or in **booster service, turbine – driven centrifugal** compressors are used. Large **integral** reciprocating compressors are also used.

Usually, the separated gas is **saturated** with water vapor and must be de-

hydrated to an acceptable level (normally less than 7 lb/MMscf). Usually this is done in a glycol dehydrator. Dry glycol is pumped to the large **vertical contact tower** where it **strips** the gas of its water vapor. The wet glycol then flows through a separator to the large horizontal reboiler where it is heated and the water boiled off as steam.

In some locations it may be necessary to remove the heavier hydrocarbons to lower the hydrocarbon **dew point**. Contaminants such as H_2S and CO_2 may be present at levels higher than those acceptable to the gas purchaser. If this is the case, then additional equipment will be necessary to "**sweeten**" the gas.

The oil and emulsion from the separators must be treated to remove water. Most oil contracts specify a maximum percent of **basic sediment** and water (BS and W) that can be in the crude. This will typically vary from 0.5% to 3% depending on location. Some refineries have a limit on salt content in the crude, which may require several stages of **dilution** with fresh water and subsequent treating to remove the water. Typical salt limits are 10 to 25 pounds of salt per thousand barrels.

The water that is produced with crude oil can be disposed of overboard in most offshore areas, or evaporated from pits in some locations onshore. Usually, it is injected into **disposal wells** or used for **waterflooding**. In any case, water from the separators must be treated to remove small quantities of produced oil. If the water is to be injected into a disposal well, facilities may be required to filter solid particles from it.

Any solids produced with the well stream must also be separated, cleaned, and disposed of in a manner that does not violate environmental criteria. Facilities may include **sedimentation basins** or tanks, **hydrocyclones**, **filters**.

The facility must provide for well testing and measurement so that gas, oil, and water production can be properly **allocated** to each well. This is necessary not only for **accounting** purposes but also to perform reservoir studies as the field is **depleted**.

Lease Automatic Custody Transfer (LACT)

In large facilities oil is typically sold through a LACT unit, which is designed to meet API Standards and whatever additional measuring and **sampling standards** are required by the crude purchaser. The value received for the crude will typically depend on its gravity, BS + W content, and volume. Therefore, the LACT unit must not only measure the volume accurately, but must continuously monitor the BS + W content and take a sufficiently representative sample so that the gravity and BS + W can be measured.

Figure 1.1.1 shows schematically the elements of a typical LACT u-nit. The crude first flows through a **strainer / gas eliminator** to protect the **meter** and to assure that there is no gas in the liquid. An automatic BS + W probe is mounted in a **vertical run**. When BS + W exceeds the sales contract quality this **probe** automatically **actuates** the **diverter valve**, which blocks the liquid from going further in the LACT unit and sends it back to the process for further treating. Some sales contracts allow for the BS + W probe to merely sound a warning so that the operators can manually take corrective action. The BS + W probe must be mounted in a vertical run if it is to get a **true reading** of the average quality of the stream.

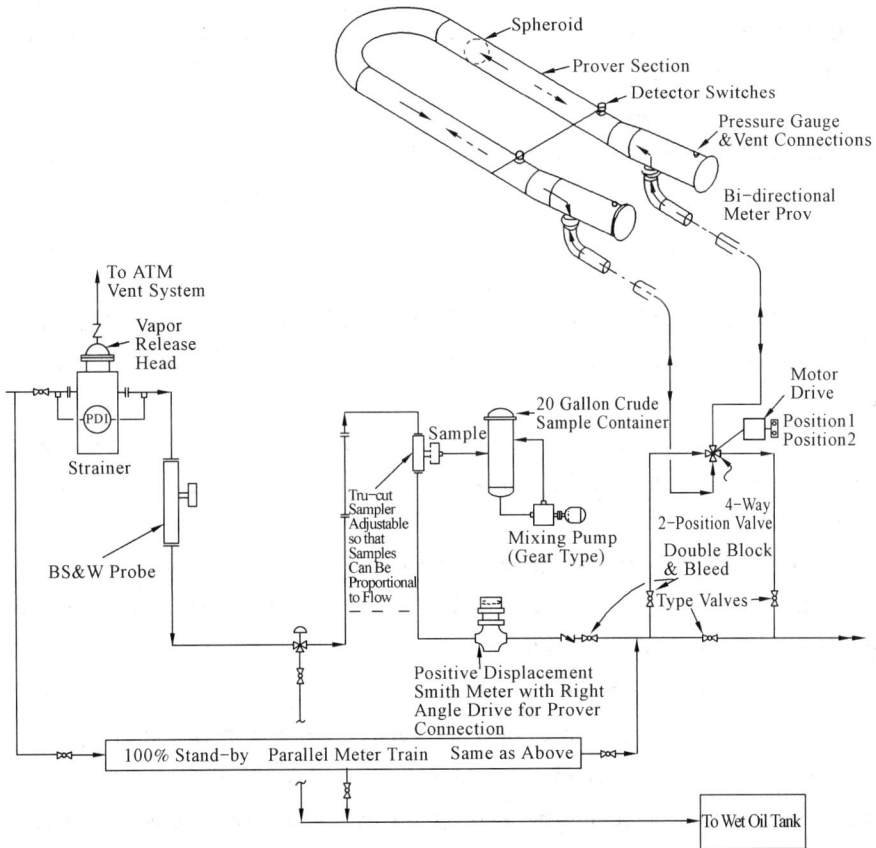

Figure 1.1.1 Typical LACT unit schematic

The **sampler** receives a signal from the meter to assure that the sample size is always proportional to flow even if the flow varies. The sample container has a mixing pump so that the liquid in the container can be mixed and made homogeneous prior to taking a sample of this fluid. It is this small sample that will be used to **convert** the meter reading for BS + W and gravity.

The liquid then flows through a **positive displacement meter**. On large installations a meter prover such as that shown in Figure 1. 1. 1 is included as a permanent part of the LACT skid or is brought to the location when a meter must be **proven**.

Gas Dehydrators

Removing most of the water vapor from the gas is required by most gas sales contracts, because it prevents **hydrates** from forming when the gas is cooled in the transmission and distribution systems and prevents water vapor from **condensing** and creating a corrosion problem. Dehydration also increases line capacity **marginally**.

Most sales contracts in the southern United States call for reducing the water content in the gas to less than 7 lb/MMscf. In colder climates, sales requirements of 3 to 5 lb/MMscf are common. The following methods can be used for drying the gas:

(1) Cool to the hydrate formation level and separate the water that forms. This can only be done where high water contents (±30 lb/MMscfd) are acceptable.

(2) Use a Low – temperature Exchange (LTX) unit designed to melt the hydrates as they are formed. LTX units require inlet pressures greater than 2500 psi to work effectively.

Although they were common in the past, they are not normally used because of their tendency to freeze and their inability to operate at lower inlet pressure as the well **FTP** declines.

(3) Contact the gas with a solid bed of $CaCl_2$. The $CaCl_2$ will reduce the moisture to low levels, but it cannot be **regenerated** and is very corrosive.

(4) Use a solid **desiccant**, such as **activated alumina**, **silica gel** or **molecular sieve**, which can be regenerated. These are relatively expensive units, but they can get the moisture content to very low levels. Therefore, they tend to be used on the inlets to low temperature gas processing plants, but are not common in production facilities.

　　(5) Use a liquid desiccant, such as **methanol** or **ethylene glycol**, which cannot be regenerated. These are relatively inexpensive. Extensive use is made of methanol to lower the hydrate temperature of gas well flowlines to keep hydrates from freezing the **choke**.

　　(6) Use a **glycol** liquid desiccant, which can be regenerated. This is the most common type of gas dehydration system.

　　Figure 1.1.2 shows how a typical **bubble − cap** glycol contact tower works. Wet gas enters the base of the tower and flows upward through the bubble caps. Dry glycol enters the top of the tower, and because of the downcomer weir on the edge of each tray, flows across the tray and down to the next. There are typically six to eight trays in most applications. The bubble caps assure that the upward flowing gas is dispersed into small bubbles to maximize its contact area with the glycol.

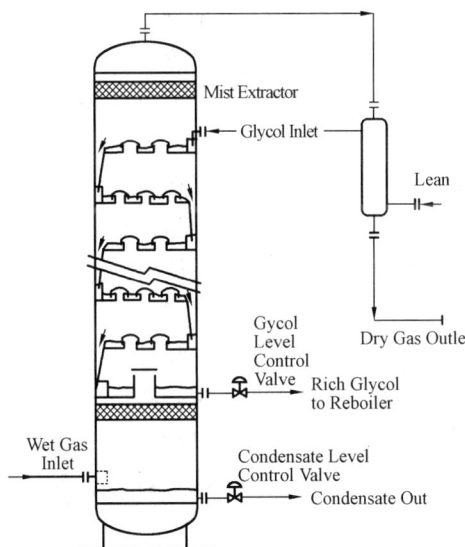

Figure 1.1.2　Typical glycol contact tower

　　Before entering the contactor the dry glycol is cooled by the outlet gas to minimize vapor losses when it enters the tower. The wet glycol leaves from the base of the tower and flows to the **reconcentrator** (reboiler) by way of **heat exchangers**, a gas separator, and filters. In the reboiler the glycol is heated to a sufficiently high temperature to drive off the water as steam. The dry glycol is then pumped back to the contact tower.

□ Words and Expressions

facility	设施;设备
flash	闪蒸
volatility	挥发度
compress	压缩
turbine – driven	燃气轮机驱动的
centrifugal	离心的
integral	整体的
saturate	饱和
dehydrate	脱水
strip	剥夺
sweeten	使甜;变舒适;净化
dilution	稀释
waterflooding	水驱
hydrocyclone	旋流分离器
filter	过滤器
allocate	配产
account	计量
deplete	枯竭
strainer	过滤器
meter	流量计
probe	探测器
actuate	启动
sampler	取样器
convert	转换
prover	标定仪
prove	标定
hydrate	水合物
condense	凝结
marginally	略微地
regenerate	再生
desiccant	干燥剂
methanol	甲醇

choke	阻塞
glycol	醇类
reconcentrator	重沸器

Phrases and Expressions

reciprocating compressor	往复式压缩机
booster service	增压作业
vertical contact tower	立式接触塔
dew point	露点
basic sediment	底部沉淀物
disposal well	污水井
sedimentation basin	沉降池
Lease Automatic Custody Transfer	矿场自动接收、取样、计量、传输系统
sampling standard	取样标准
gas eliminator	天然气逸出器
vertical run	垂向流动
diverter valve	分流阀
true reading	真实读数
positive displacement meter	正排流量计
FTP	自喷油压
activated alumina	活性矾土
silica gel	硅胶
molecular sieve	分子筛
ethylene glycol	乙二醇
bubble – cap glycol	泡罩
heat exchanger	换热器

Language Focus

1. The job of a production facility is to separate the well stream into three components, typically called "phases" (oil, gas, and water), and process these phases into some marketable product(s) or dispose of them in an environmentally acceptable manner.

(参考译文:采油设施将从井下采出的液体分离成三种成分(即通常所说的原油、天然气和水三相),并将该三相流体分别加工成为市场销售产品,或将其处理成为环境保护条例所允许的排放物。)

本句中不定式放在"be"后面形成表语,表示具体的工作内容;"called"是过去分词短语作定语修饰"three components"。

2. Dry glycol is pumped to the large vertical contact tower where it strips the gas of its water vapor.

(参考译文:干乙二醇用泵送至大型立式接触塔内,在该塔内将天然气中的水蒸气吸收。)

本句中"where"引导限制性定语从句,相当于"in the large vertical contact tower";"it"指代"dry glycol"。

3. The wet glycol then flows through a separator to the large horizontal reboiler where it is heated and the water boiled off as steam.

(参考译文:湿乙二醇从分离器中流出至大型卧式重沸器内,在重沸器中加热,并蒸发掉水分。)

本句中"where"引导限制性定语从句,相当于"in the large horizontal reboiler";"it"指代"the wet glycol",翻译时可以省略。

4. Some refineries have a limit on salt content in the crude, which may require several stages of dilution with fresh water and subsequent treating to remove the water.

(参考译文:某些炼油厂对原油中的含盐量作出限定。这样,就需要用淡水对原油进行数级脱盐处理。)

本句中"which"引导非限制性定语从句,代替前面的句子。

5. This is necessary not only for accounting purposes but also to perform reservoir studies as the field is depleted.

(参考译文:这样做不仅是为了计量,也是为了在油田行将枯竭之时对油层进行评价研究。)

not only… but also… 为固定搭配,并列两部分目的,not only 后的目的是由 for 介词短语表示的,but also 后的目的是由不定式 to 表示的;as 引导时间状语从句。

□ Reinforced Learning

I. Answer the following questions for a comprehension of the text.

1. What's the job of a production facility?

2. What should we do to treat contaminants such as H_2S and CO_2 at high levels?

3. Why must the facility provide for well testing and measurement?

4. How does an automatic BS + W probe function?

5. What do most sales contracts in the southern United States call for?

Ⅱ. Multiple choice：choose the correct one from the alternative answers to give the exact meaning of the words.

1. These steps remove enough light hydrocarbons to produce a stable crude oil with the volatility to meet sales criteria.

　　A. erratum　　　B. futility　　　C. motility　　　D. vapor pressure

2. The gas that is separated must be compressed and treated for sales.

　　A. compact　　　B. impressed　　　C. opposed　　　D. unpressed

3. In large facilities or in booster service, turbine – driven centrifugal compressors are used.

　　A. central　　　B. concentric　　　C. eccentric　　　D. homocentric

4. Large integral reciprocating compressors are also used.

　　A. integrated　　B. internal　　　C. integated　　　D. interval

5. Usually, the separated gas is saturated with water vapor and must be dehydrated to an acceptable level.

　　A. involved　　　B. filled　　　C. penetrated　　　D. scattered

6. Dry glycol is pumped to the large vertical contact tower where it strips the gas of its water vapor.

　　A. removes　　　B. stripes　　　C. wipes　　　D. splits

7. The facility must provide for well testing and measurement so that gas, oil, and water production can be properly allocated to each well.

　　A. switched　　　B. performed　　　C. distributed　　　D. manufactured

8. This is necessary not only for accounting purposes but also to perform reservoir studies as the field is depleted.

　　A. explored　　　B. exploited　　　C. excavated　　　D. exhausted

9. Dehydration also increases line capacity marginally.

　　A. greatly　　　B. slightly　　　C. remarkably　　　D. significantly

10. The $CaCl_2$ will reduce the moisture to low levels, but it cannot be regenerated and is very corrosive.

　　A. generated　　　B. recovered　　　C. renewed　　　D. reduced

III. Multiple choice: read the four suggested translations and choose the best answer.

1. In large facilities or in booster service, turbine – driven centrifugal compressors are used.

   A. 增强　　　　B. 增压　　　　C. 增重　　　　D. 增速

2. Usually, the separated gas is saturated with water vapor and must be dehydrated to an acceptable level.

   A. 充满　　　　B. 充斥　　　　C. 饱和　　　　D. 负荷

3. Usually, the separated gas is saturated with water vapor and must be dehydrated to an acceptable level.

   A. 脱硫　　　　B. 脱水　　　　C. 脱氢　　　　D. 脱氮

4. If this is the case, then additional equipment will be necessary to "sweeten" the gas.

   A. 甜化　　　　B. 淡化　　　　C. 液化　　　　D. 净化

5. Some refineries have a limit on salt content in the crude, which may require several stages of dilution with fresh water and subsequent treating to remove the water.

   A. 稀释　　　　B. 调和　　　　C. 调适　　　　D. 溶解

6. The facility must provide for well testing and measurement so that gas, oil, and water production can be properly allocated to each well.

   A. 分离　　　　B. 分散　　　　C. 配备　　　　D. 配产

7. This is necessary not only for accounting purposes but also to perform reservoir studies as the field is depleted.

   A. 计量　　　　B. 计算　　　　C. 算术　　　　D. 测量

8. In large facilities oil is typically sold through a LACT unit, which is designed to meet API Standards and whatever additional measuring and sampling standards are required by the crude purchaser.

   A. 设施　　　　B. 机器　　　　C. 采油厂　　　　D. 公司

9. When BS + W exceeds the sales contract quality this probe automatically actuates the diverter valve, which blocks the liquid from going further in the LACT unit and sends it back to the process for further treating.

   A. 启动　　　　B. 联系　　　　C. 关闭　　　　D. 激发

10. The wet glycol leaves from the base of the tower and flows to the reconcentrator by way of heat exchangers, a gas separator, and filters.

   A. 逸出器　　　　B. 探测器　　　　C. 重沸器　　　　D. 换热器

IV. Put the following sentences into Chinese.

1. The job of a production facility is to separate the well stream into three components, typically called "phases" (oil, gas, and water), and process these phases into some marketable product(s) or dispose of them in an environmentally acceptable manner.

2. Separators can be either horizontal or vertical in configuration.

3. In large facilities oil is typically sold through a LACT unit, which is designed to meet API Standards and whatever additional measuring and sampling standards are required by the crude purchaser.

4. The sampler receives a signal from the meter to assure that the sample size is always proportional to flow even if the flow varies.

5. Although they were common in the past, they are not normally used because of their tendency to freeze and their inability to operate at lower inlet pressure as the well FTP declines.

V. Put the following paragraphs into Chinese.

1. The water that is produced with crude oil can be disposed of overboard in most offshore areas, or evaporated from pits in some locations onshore. Usually, it is injected into disposal wells or used for waterflooding. In any case, water from the separators must be treated to remove small quantities of produced oil. If the water is to be injected into a disposal well, facilities may be required to filter solid particles from it.

2. Removing most of the water vapor from the gas is required by most gas sales contracts, because it prevents hydrates from forming when the gas is cooled in the transmission and distribution systems and prevents water vapor from condensing and creating a corrosion problem. Dehydration also increases line capacity marginally.

1.2 Basic System Configuration

Guidance to Reading

In oil production, since flows from two or more wells are usually commingled in a central facility, it is necessary to install a manifold to allow flow from any one well to be produced into any of the bulk or test production sys-

tems. Multistage separation is used to allow controlled separation of volatile components. The purpose is to achieve maximum liquid recovery and stabilized oil and gas, and separate water. There also has to be a certain minimum pressure difference between each stage to allow satisfactory performance in the pressure and level control loops.

🔲 **Text**

Wellhead and Manifold

The production system begins at the wellhead, which should include at least one **choke**, unless the well is on **artificial lift**. Most of the pressure drop between the well **flowing tubing pressure** (FTP) and the initial separator operating pressure occurs across this choke. The size of the opening in the choke determines the flow rate, because the pressure upstream is determined primarily by the well FTP, and the pressure downstream is determined primarily by the pressure control valve on the first separator in the process. For high – pressure wells it is desirable to have a **positive choke in series with** an adjustable choke. The positive choke **takes over** and keeps the production rate within limits should the adjustable choke fail.

On offshore facilities and other high – risk situations, an automatic shutdown valve should be installed on the wellhead (It is required by federal law in the United States). In all cases, **block valves** are needed so that maintenance can be performed on the choke if there is a long flowline.

Whenever flows from two or more wells are commingled in a central facility, it is necessary to install a manifold to allow flow from any one well to be produced into any of the bulk or test production systems.

Initial Separator Pressure

Because of the **multicomponent** nature of the produced fluid, the higher the pressure at which the initial separation occurs, the more liquid will be obtained in the separator. This liquid contains some light components that vaporize in the stock tank downstream of the separator.

If the pressure for initial separation is too high, too many light components will stay in the liquid phase at the separator and be lost to the gas phase at the tank. If the pressure is too low, not as many of these light components will be stabilized into the liquid at the separator and they will be lost to the gas phase.

Stage Separation

In a simple **single – stage** process, the fluids are flashed in an initial separator and then the liquids from that separator are flashed again at the stock tank. Traditionally, the stock tank is not normally considered a separate stage of separation, though it most assuredly is. Figure 1.2.1 shows a three – stage separation process. The liquid is first flashed at an initial pressure and then flashed at **successively** lower pressures two times before entering the stock tank.

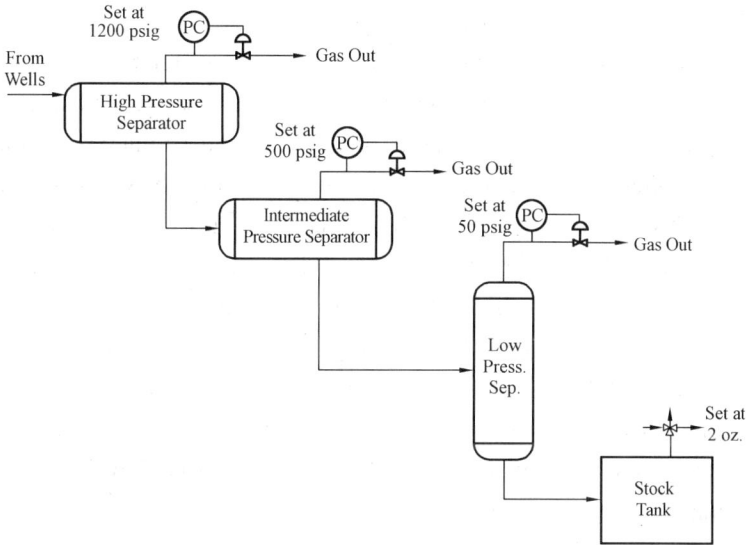

Figure 1.2.1   Stage separation

Because of the multicomponent nature of the produced fluid, it can be shown by flash calculations that the more stages of separation after the initial separation the more light components will be stabilized into the liquid phase. This can be understood qualitatively by realizing that in a stage separation process the light hydrocarbon molecules that flash are removed at relatively high pressure, keeping the **partial pressure** of the intermediate hydrocarbons lower at each stage. As the number of stages approaches **infinity**, the lighter molecules are removed as soon as they are formed and the partial pressure of the **intermediate components** is maximized at each stage. The compressor horsepower required is also reduced by stage separation as some of the gas is captured at a higher pressure than would otherwise have occurred.

Selection of Stages

As more stages are added to the process there is less and less **incremental** liquid recovery. The **diminishing** income for adding a stage must more than offset the cost of the additional separator, piping, controls, space, and compressor complexities. It is clear that for each facility there is an **optimum** number of stages. In most cases, the optimum number of stages is very difficult to determine as it may be different from well to well and it may change as the wells′flowing pressure declines with time.

Fields with Different Flowing Tubing Pressures

The discussion to this point has focused on a situation where all the wells in a field produce at roughly the same flowing tubing pressure, and stage separation is used to maximize liquid production and minimize compressor horsepower. Often, stage separation is used because different wells producing to the facility have different flowing tubing pressures. This could be because they are completed in different reservoirs, or are located in the same reservoir but have different water production rates. By using a manifold arrangement and different primary separator operating pressures, there is not only the benefit of stage separation of high – pressure liquids, but also **conservation** of reservoir energy. High – pressure wells can continue to flow at sales pressure requiring no compression, while those with lower tubing pressures can flow into whichever system minimizes compression.

Two Phase vs. Three Phase Separators

In our example process the high – and intermediate – stage separators are two – phase, while the low – pressure separator is three – phase. This is called a "free water knockout" (FWKO) because it is designed to separate the free water from the oil and emulsion, as well as separate gas from liquid. The choice depends on the expected flowing characteristics of the wells. If large amounts of water are expected with the high – pressure wells, it is possible that the size of the other separators could be reduced if the high – pressure separator was three – phase.

Oil Treating

Most oil treating on offshore facilities is done in vertical or horizontal treaters. Figure 1. 2. 2 is an oil treater. In this case, a **gas blanket** is provided to assure that there is always enough pressure in the treater so the water will flow to water treating.

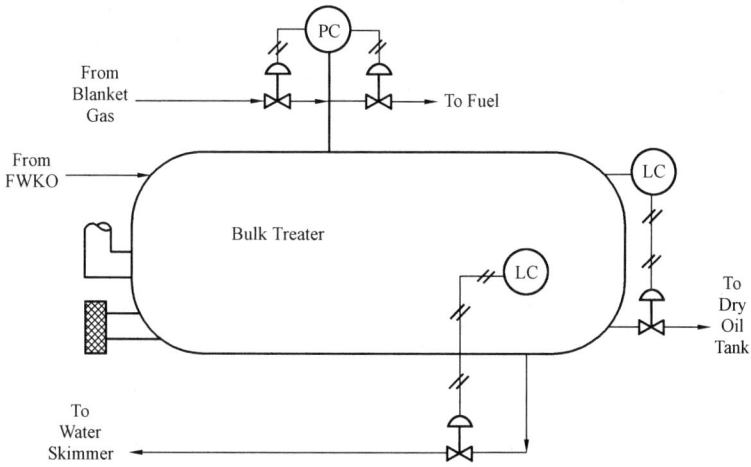

Figure 1.2.2　Bulk treater

At onshore locations the oil may be treated in a big "**gunbarrel**" (or settling) tank, as shown in Figure 1.2.3. All tanks should have a pressure/vacuum valve with **flame arrester** and gas blanket to keep a positive pressure on the system and **exclude** oxygen. This helps to prevent corrosion, eliminate a **potential safety hazard**, and conserve some of the hydrocarbon vapors. Figure 1.2.4 shows a typical **pressure/vacuum valve**. A pressure in the tank lifts a **weighted disk**, which allows the gas to escape. If there is a vacuum in the tank because

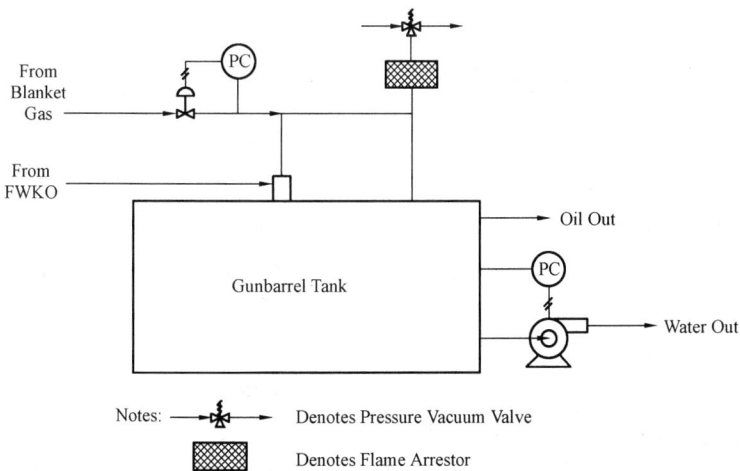

Figure 1.2.3　Gunbarrel

the gas blanket failed to maintain a slight positive pressure, the greater **ambient** pressure lifts another disk, which allows air to enter. Although we wish to exclude air, it is preferable to allow a small controlled volume into the tank rather than allow the tank to **collapse**.

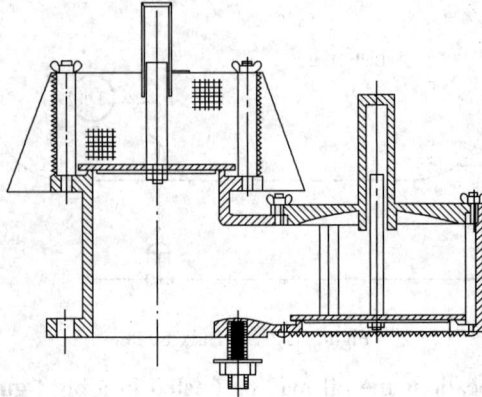

Figure 1.2.4　Typical pressure/vacuum valve

The oil is skimmed off the surface of the gun barrel and the water exits from the bottom either through a **water leg** or an interface controller and dump valve. It must be pointed out that since the volume of the liquid is fixed by the oil outlet, gunbarrels cannot be used as surge tanks.

Flow from the treater or gun barrel goes to a **surge tank** from which it either flows into a barge or truck or is pumped into a pipeline.

Words and Expressions

choke	油嘴
multicomponent	多组分
single – stage	单级
successively	连续地
infinity	无限
incremental	增量的
diminishing	递减的
optimum	最佳
conservation	保存
gunbarrel	沉降罐

exclude	脱除
ambient	周围的
collapse	倒塌,压扁

Phrases and Expressions

artificial lift	人工举升
flowing tubing pressure	自喷油压
positive choke	固定油嘴
in series with	串联
take over	取代
block valve	截断阀
initial separator	初级分离器
stage separation	多级分离
partial pressure	分压
intermediate component	中间组分
gas blanket	气体覆盖层
flame arrester	阻火器
potential safety hazard	安全隐患
pressure/vacuum valve	压力真空阀
weighted disk	加重盘
water leg	含水区
surge tanks	缓冲罐

Language Focus

1. The size of the opening in the choke determines the flow rate, because the pressure upstream is determined primarily by the well FTP, and the pressure downstream is determined primarily by the pressure control valve on the first separator in the process.

（参考译文：油嘴的开孔直径决定了油流流量,因为油流的上游压力主要是取决于油井的自喷油压,而下游压力主要是取决于采油过程中第一个分离器上的压力控制阀的压力大小。）

本句中"because"引导原因状语从句;从句中"and"连接结构相同的两个并列分句,都是主语加谓语"is determined primarily by"。

2. Because of the multicomponent nature of the produced fluid, the higher

the pressure at which the initial separation occurs, the more liquid will be obtained in the separator.

（参考译文：由于采出的流体具有多组分特性，因此，初次分离过程中油流压力越高，则分离器中所能获得的液流就越多。）

本句中"Because of the multicomponent nature of the produced fluid"是原因状语；主句用了两个"定冠词加比较级"的结构"the higher..., the more..."表示"越……越……"；"at which"引导限制性定语从句，指代"at the pressure"，先行词是"pressure"。

3. This can be understood qualitatively by realizing that in a stage separation process the light hydrocarbon molecules that flash are removed at relatively high pressure, keeping the partial pressure of the intermediate hydrocarbons lower at each stage.

（参考译文：也就是说，在多级分离的工艺条件下，能闪蒸的轻烃分子在相当高压条件下会分离出来，使每级分离过程中的中间烃类分压处于较低状态。）

本句中"This can be understood qualitatively by realizing..."表示"可以通过认识到……明确这个特性"，是对前一句话的补充说明；后面的"that"引导宾语从句，第二个"that"引导限制性定语从句，先行词是"molecules"，相当于"the light hydrocarbon molecules"；"keeping the partial pressure of the intermediate hydrocarbons lower at each stage"做伴随状语。

4. In our example process the high - and intermediate - stage separators are two - phase, while the low - pressure separator is three - phase.

（参考译文：在我们所举实例中，高压及中等压力等级的分离器均是两相分离器，而低压分离器则是三相分离器。）

本句中"while"表示对比，"the high - and intermediate - stage separators"与"the low - pressure separator"对比。

5. If there is a vacuum in the tank because the gas blanket failed to maintain a slight positive pressure, the greater ambient pressure lifts another disk, which allows air to enter.

（参考译文：若由于罐内某一层气罩失效，不能维持稍高正压而出现真空，则外部更高的压力会顶起另一圆盘，使空气进入。）

"If"引导条件状语从句，从句中有"because"引导的原因状语从句，"which"引导非限定性状语从句，指代主句"the greater ambient pressure lifts another disk"。

Reinforced Learning

I. Answer the following questions for a comprehension of the text.

1. What does a wellhead have?

2. How does pressure influence the light components in initial separation?

3. How to select stages?

4. Which one should we choose, Two Phase Separators or Three Phase Separators?

5. How shall we treat the oil at onshore locations?

II. Multiple choice: choose the correct one from the alternative answers to give the exact meaning of the words.

1. Most of the pressure drop between the well flowing tubing pressure and the initial separator operating pressure occurs across this choke.

A. aboard B. over C. through D. by

2. The positive choke takes over and keeps the production rate within limits should the adjustable choke fail.

A. changes B. replaces C. increases D. loses

3. As the number of stages approaches infinity, the lighter molecules are removed as soon as they are formed and the partial pressure of the intermediate components is maximized at each stage.

A. magnitude B. limit C. minimum D. boundary

4. As more stages are added to the process there is less and less incremental liquid recovery.

A. incredible B. interval C. infinite D. increasing

5. The diminishing income for adding a stage must more than offset the cost of the additional separator, piping, controls, space, and compressor complexities.

A. reducing B. increasing C. extinguishing D. expanding

6. It is clear that for each facility there is an optimum number of stages.

A. optimistic B. excellent C. outstanding D. optimal

7. By using a manifold arrangement and different primary separator operating pressures, there is not only the benefit of stage separation of high – pressure liquids, but also conservation of reservoir energy.

A. construction   B. protection

C. consumption   D. conservativeness

8. At onshore locations the oil may be treated in a big "gunbarrel", as shown in Figure 1. 2. 3.

A. rifle   B. truck   C. barrel   D. tank

9. All tanks should have a pressure/vacuum valve with flame arrester and gas blanket to keep a positive pressure on the system and exclude oxygen.

A. include   B. keep   C. prevent   D. absorb

10. Although we wish to exclude air, it is preferable to allow a small controlled volume into the tank rather than allow the tank to collapse.

A. crash   B. fail   C. remain   D. stand

III. Multiple choice:read the four suggested translations and choose the best answer.

1. The production system begins at the wellhead, which should include at least one choke, unless the well is on artificial lift.

A. 油帽   B. 油嘴   C. 油杆   D. 油井

2. For high-pressure wells it is desirable to have a positive choke in series with an adjustable choke.

A. 关联   B. 连串   C. 联系   D. 串联

3. The positive choke takes over and keeps the production rate within limits should the adjustable choke fail.

A. 更换   B. 拿来   C. 取代   D. 去除

4. The liquid is first flashed at an initial pressure and then flashed at successively lower pressures two times before entering the stock tank.

A. 更加   B. 继续   C. 连续   D. 接着

5. As the number of stages approaches infinity, the lighter molecules are removed as soon as they are formed and the partial pressure of the intermediate components is maximized at each stage.

A. 无限   B. 有限   C. 无界   D. 有界

6. It is clear that for each facility there is an optimum number of stages.

A. 良好   B. 优秀   C. 最佳   D. 最大

7. All tanks should have a pressure/vacuum valve with flame arrester and gas blanket to keep a positive pressure on the system and exclude oxygen.

A. 阻力器   B. 阻阀器   C. 阻火器   D. 阻喷器

8. This helps to prevent corrosion, eliminate a potential <u>safety hazard</u>, and conserve some of the hydrocarbon vapors.

A. 安全灾难　　　B. 安全灾害　　　C. 安全问题　　　D. 安全隐患

9. If there is a vacuum in the tank because the gas blanket failed to maintain a slight positive pressure, the greater <u>ambient</u> pressure lifts another disk, which allows air to enter.

A. 围绕的　　　B. 周围的　　　C. 上下的　　　D. 反向的

10. It must be pointed out that since the volume of the liquid is fixed by the oil outlet, gun barrels cannot be used as <u>surge tanks</u>.

A. 解压罐　　　B. 缓冲罐　　　C. 储存罐　　　D. 沉降罐

IV. Put the following sentences into Chinese.

1. The size of the opening in the choke determines the flow rate, because the pressure upstream is determined primarily by the well FTP, and the pressure downstream is determined primarily by the pressure control valve on the first separator in the process.

2. The positive choke takes over and keeps the production rate within limits should the adjustable choke fail.

3. Whenever flows from two or more wells are commingled in a central facility, it is necessary to install a manifold to allow flow from any one well to be produced into any of the bulk or test production systems.

4. Because of the multicomponent nature of the produced fluid, it can be shown by flash calculations that the more stages of separation after the initial separation the more light components will be stabilized into the liquid phase.

5. By using a manifold arrangement and different primary separator operating pressures, there is not only the benefit of stage separation of high - pressure liquids, but also conservation of reservoir energy.

V. Put the following paragraphs into Chinese.

1. As more stages are added to the process there is less and less incremental liquid recovery. The diminishing income for adding a stage must more than offset the cost of the additional separator, piping, controls, space, and compressor complexities. It is clear that for each facility there is an optimum number of stages. In most cases, the optimum number of stages is very difficult to determine as it may be different from well to well and it may change as the wells' flowing

pressure declines with time.

2. Often, stage separation is used because different wells producing to the facility have different flowing tubing pressures. This could be because they are completed in different reservoirs, or are located in the same reservoir but have different water production rates. By using a manifold arrangement and different primary separator operating pressures, there is not only the benefit of stage separation of high – pressure liquids, but also conservation of reservoir energy. High – pressure wells can continue to flow at sales pressure requiring no compression, while those with lower tubing pressures can flow into whichever system minimizes compression.

1.3   Two – Phase Oil and Gas Separation

卜 Guidance to Reading

A vessel designed to separate production fluids into their constituent components of oil, gas and water is a separator referred to in the following ways: separator, trap, knockout vessel, flash chamber, scrubber and filter. These separating vessels are normally used on a producing lease or platform near the wellhead, or tank battery to separate fluids produced from oil and gas wells into oil and gas or liquid and gas. This article deals with two – phase separators. In addition, it discusses the requirements of good separation design and how various mechanical devices take advantage of the physical forces in the produced stream to achieve good separation.

卜 Text

Produced wellhead fluids are complex mixtures of different compounds of hydrogen and carbon, all with different densities, vapor pressures, and other physical characteristics. As a well stream flows from the hot, high – pressure petroleum reservoir, it experiences pressure and temperature reductions. Gases **evolve** from the liquids and the well stream changes in character. The velocity of the gas carries liquid droplets, and the liquid carries gas bubbles. The **physical separation** of these phases is one of the basic operations in the production, processing, and treatment oil and gas.

In oil and gas separator design, we mechanically separate from a hydrocarbon stream the liquid and gas components that exist at a specific temperature and pressure. Proper separator design is important because a separation vessel is normally the initial processing vessel in any facility, and improper design of this process component can "bottleneck" and reduce the capacity of the entire facility.

Separators are classified as "**two – phase**" if they separate gas from the total liquid stream and "**three – phase**" if they also separate the liquid stream into its crude oil and water **components**. Separators are sometimes called "**gas scrubbers**" when the ratio of gas rate to liquid rate is very high. Some operators use the term "**traps**" to designate separators that handle flow directly from wells. In any case, they all have the same configuration and are sized **in accordance with** the same procedure.

Horizontal Separators

Separators are designed in either horizontal, vertical, or spherical **configurations**. Figure 1.3.1 is a schematic of a horizontal separator. The fluid enters the separator and hits an inlet diverter causing a sudden change in **momentum**. The initial gross separation of liquid and vapor occurs at the **inlet diverter**. The force of gravity causes the liquid droplets to fall out of the gas stream to the bottom of the vessel where it is collected. This liquid collection section provides the **retention time** required to let entrained gas evolve out of the oil and rise to the vapor space. It also provides a **surge volume**, if necessary, to handle **intermittent slugs of liquid**. The liquid then leaves the vessel through the **liquid dump valve**. The liquid dump valve is regulated by a level controller. The level controller senses changes in liquid level and controls the dump valve accordingly.

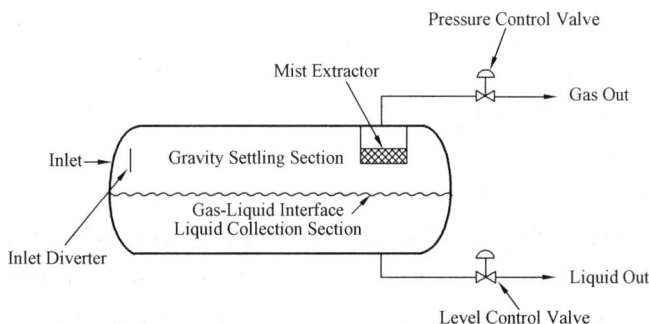

Figure 1.3.1 Horizontal separator schematic

The gas flows over the inlet diverter and then horizontally through the gravity settling section above the liquid. As the gas flows through this section, small drops of liquid that were **entrained** in the gas and not separated by the inlet diverter are separated out by gravity and fall to the gas – liquid interface.

Some of the drops are of such a small diameter that they are not easily separated in the gravity settling section. Before the gas leaves the vessel it passes through a **coalescing section** or mist extractor. This section uses elements of **vanes**, **wire mesh**, or plates to coalesce and remove the very small droplets of liquid in one final separation before the gas leaves the vessel.

The pressure in the separator is maintained by a pressure controller. The pressure **controller senses** changes in the pressure in the separator and sends a signal to either open or close the pressure control valve accordingly. By controlling the rate at which gas leaves the vapor space of the vessel the pressure in the vessel is maintained. Normally, horizontal separators are operated half full of liquid to maximize the surface area of the gas liquid interface.

Vertical separators

Figure 1.3.2 is a schematic of a vertical separator. In this configuration the inlet flow enters the vessel through the side. As in the horizontal separator, the inlet diverter does the initial gross separation. The liquid flows down to the liquid collection section of the vessel. Liquid continues to flow downward through this section to the liquid outlet. As the liquid reaches **equilibrium**, gas bubbles flow counter to the direction of the liquid flow and eventually migrate to the vapor space. The level controller and liquid dump valve operate the same as in a horizontal separator.

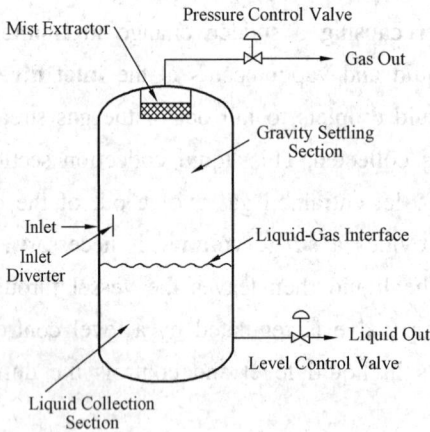

Figure 1.3.2 Vertical separator schematic

The gas flows over the inlet diverter and then vertically upward toward the gas outlet. In the **gravity settling section** the liquid drops fall vertically downward **counter to** the gas flow. Gas goes through the **mist extractor** section before it leaves the vessel. Pressure and level are maintained as in a horizontal separator.

Spherical Separators

A typical spherical separator is shown in Figure 1. 3. 3. The same four sections can be found in this vessel. Spherical separators are a special case of a vertical separator where there is no **cylindrical shell** between the two heads. They may be very efficient from a **pressure containment** standpoint but because (1)they have limited liquid surge capability and (2)they exhibit **fabrication** difficulties, they are not usually used in oil field facilities. For this reason we will not be discussing spherical separators any further.

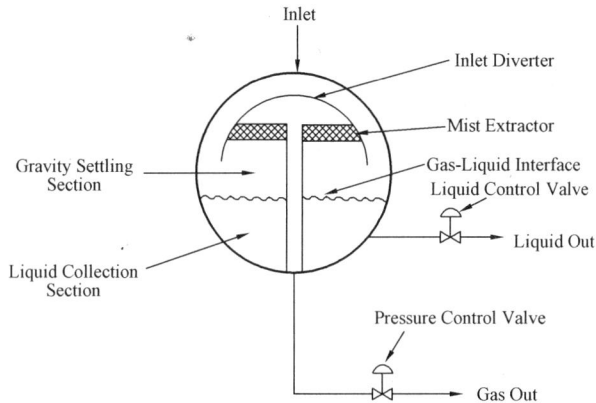

Figure 1. 3. 3   Spherical separator schematic

Other Configurations

Cyclone separators are designed to operate by **centrifugal force**. These designs are best suited for fairly clean gas streams. The swirling action of the gas stream as it enters the scrubber separates the droplets and dust from the gas stream by centrifugal force. Although such designs can result in significantly smaller sizes, they are not commonly used in production operations because (1)their design is rather sensitive to **flow rate** and (2) they require greater pressure drop than the standard configurations previously described. Since separation efficiency decreases as velocity decreases, cyclone separators are not suitable for widely varying flow rates. These units are commonly used to recover glycol carryover downstream of a **dehydration tower**. In recent years, demand for using cyclone separators on floating facilities has increased because space and weight considerations are overriding on such facilities.

Two – barrel separators are common where there is a very low liquid flow rate. In these designs the gas and liquid chambers are separated. Another type of separator that is frequently used in some high – gas/low – liquid flow applications is a **filter separator**. These can be either horizontal or vertical in configuration.

Words and Expressions

evolve	放出
two – phase	两相
three – phase	三相
component	组分
trap	捕集器
configuration	结构
momentum	动能
entrained	夹带的
vane	叶片
equilibrium	平衡
fabrication	制造

Phrases and Expressions

physical separation	物理分离
gas scrubber	气体洗涤器
in accordance with	根据
inlet diverter	入口分流器
retention time	保留时间
surge volume	缓冲容积
intermittent slugs of liquid	间歇性水击现象
liquid dump valve	液体放空阀
coalescing section	聚结段
wire mesh	金属丝网
controller sense	传感
gravity settling section	重力沉降段
counter to	逆向
mist extractor	除雾器
spherical separator	球形分离器
cylindrical shell	圆筒壳

pressure containment	承压
cyclone separator	旋转分离器
centrifugal force	离心力
dehydration tower	脱水塔
two – barrel separator	双筒式分离器
flow rate	流量
filter separator	过滤分离器

Language Focus

1. As a well stream flows from the hot, high – pressure petroleum reservoir, it experiences pressure and temperature reductions.

（参考译文：当井下流体从高温、高压的油层中流出时，温度和压力都会降低。）

本句中"as"引导时间状语从句"a well stream flows from the hot, high – pressure petroleum reservoir"。

2. As the gas flows through this section, small drops of liquid that were entrained in the gas and not separated by the inlet diverter are separated out by gravity and fall to the gas – liquid interface.

（参考译文：当天然气进入并通过此空间时，天然气中携带的虽然经过入口分流器但仍未全部分离出来的小雾滴就会靠重力而落至气液界面处。）

本句中"as"引导时间状语从句"the gas flows through this section"；"that"引导限制性定语从句，先行词是"drops"，指代"small drops of liquid"。

3. Although such designs can result in significantly smaller sizes, they are not commonly used in production operations because（1）their design is rather sensitive to flow rate and（2）they require greater pressure drop than the standard configurations previously described.

（参考译文：虽然这种结构的分离器尺寸较小，但在采油作业中很少使用，这是因为：（1）这种分离器的设计对流量非常敏感；（2）这种分离器比前述常规分离器需要较大的压降。）

本句中"although"引导让步状语从句"such designs can result in significantly smaller sizes"；"because"引导原因状语从句"（1）their design is rather sensitive to flow rate and（2）they require greater pressure drop than the standard configurations previously described"；"previously described"作后置定语修饰"the standard configurations"。

🗗 Reinforced Learning

I. Answer the following questions for a comprehension of the text.

1. Why is proper separator design important?

2. What are two – phase separators and three – phase separators?

3. What are three designs of separators?

4. How is the pressure maintained in the separator?

5. Why are cyclone separators not commonly used in production operations?

II. Multiple choice: choose the correct one from the alternative answers to give the exact meaning of the words.

1. Gases evolve from the liquids and the well stream changes in character.

A. grow   B. emit   C. change   D. develop

2. In oil and gas separator design, we mechanically separate from a hydrocarbon stream the liquid and gas components that exist at a specific temperature and pressure.

A. counterparts   B. features   C. constituents   D. fractions

3. In addition, it discusses the requirements of good separation design and how various mechanical devices take advantage of the physical forces in the produced stream to achieve good separation.

A. utilize   B. operate   C. control   D. execute

4. In any case, they all have the same configuration and are sized in accordance with the same procedure.

A. architecture   B. structure   C. fraction   D. portion

5. In any case, they all have the same configuration and are sized in accordance with the same procedure.

A. in comparison with   B. in contrast to

C. as a consequence of   D. in conformity to

6. The fluid enters the separator and hits an inlet diverter causing a sudden change in momentum.

A. speed   B. action   C. impetus   D. thrust

7. As the liquid reaches equilibrium, gas bubbles flow counter to the direction of the liquid flow and eventually migrate to the vapor space.

A. balance　　　B. equation　　　C. equity　　　D. maximum

8. As the liquid reaches equilibrium, gas bubbles flow <u>counter to</u> the direction of the liquid flow and eventually migrate to the vapor space.

A. next to　　　B. along with　　C. according to　　D. opposite to

9. Spherical separators are a special case of a vertical separator where there is no <u>cylindrical</u> shell between the two heads.

A. round　　　B. triangular　　C. tubelike　　　D. rectangular

10. They may be very efficient from a pressure containment standpoint but because (1) they have limited liquid surge capability and (2) they exhibit <u>fabrication</u> difficulties, they are not usually used in oil field facilities.

A. exploitation　　B. production　　C. drilling　　　D. transportation

III. Multiple choice：read the four suggested translations and choose the best answer.

1. In oil and gas separator design, we mechanically separate from a hydrocarbon stream the liquid and gas <u>components</u> that exist at a specific temperature and pressure.

A. 组分　　　B. 部分　　　C. 元素　　　　D. 因素

2. The fluid enters the separator and hits an inlet diverter causing a sudden change in <u>momentum</u>.

A. 动能　　　B. 动量　　　C. 能量　　　　D. 冲动

3. It also provides a surge <u>volume</u>, if necessary, to handle intermittent slugs of liquid.

A. 数量　　　B. 大小　　　C. 卷集　　　　D. 容积

4. Before the gas leaves the vessel it passes through a <u>coalescing section</u> or mist extractor.

A. 聚集段　　B. 聚结段　　C. 积存部　　　D. 存储部

5. The pressure <u>controller senses</u> changes in the pressure in the separator and sends a signal to either open or close the pressure control valve accordingly.

A. 控感　　　B. 传感　　　C. 感觉　　　　D. 情绪

6. As the liquid reaches <u>equilibrium</u>, gas bubbles flow counter to the direction of the liquid flow and eventually migrate to the vapor space.

A. 平等　　　B. 一致　　　C. 平衡　　　　D. 均等

7. They may be very efficient from a <u>pressure containment</u> standpoint but

because (1) they have limited liquid surge capability and (2) they exhibit fabrication difficulties, they are not usually used in oil field facilities.

  A. 限压   B. 压缩   C. 承压   D. 压迫

  8. They may be very efficient from a pressure containment standpoint but because (1) they have limited liquid surge capability and (2) they exhibit fabrication difficulties, they are not usually used in oil field facilities.

  A. 锻制   B. 炼制   C. 制订   D. 制造

  9. Cyclone separators are designed to operate by centrifugal force.

  A. 向心力 B. 能动力 C. 旋转力   D. 离心力

  10. Although such designs can result in significantly smaller sizes, they are not commonly used in production operations because (1) their design is rather sensitive to flow rate and (2) they require greater pressure drop than the standard configurations previously described.

  A. 流量   B. 流速   C. 流率   D. 流失

Ⅳ. Put the following sentences into Chinese.

  1. Produced wellhead fluids are complex mixtures of different compounds of hydrogen and carbon, all with different densities, vapor pressures, and other physical characteristics.

  2. Proper separator design is important because a separation vessel is normally the initial processing vessel in any facility, and improper design of this process component can "bottleneck" and reduce the capacity of the entire facility.

  3. By controlling the rate at which gas leaves the vapor space of the vessel the pressure in the vessel is maintained.

  4. The swirling action of the gas stream as it enters the scrubber separates the droplets and dust from the gas stream by centrifugal force.

  5. Although such designs can result in significantly smaller sizes, they are not commonly used in production operations because (1) their design is rather sensitive to flow rate and (2) they require greater pressure drop than the standard configurations previously described.

Ⅴ. Put the following paragraphs into Chinese.

  1. This article deals with two – phase separators. In addition, it discusses the requirements of good separation design and how various mechanical devices take advantage of the physical forces in the produced stream to achieve good

separation.

2. Separators are classified as "two – phase" if they separate gas the total liquid stream and "three – phase" if they also separate the liquid stream into its crude oil and water components.

3. As in the horizontal separator, the inlet diverter does the initial gross separation. The liquid flows down to the liquid collection section of the vessel. Liquid continues to flow downward through this section to the liquid outlet. As the liquid reaches equilibrium, gas bubbles flow counter to the direction of the liquid flow and eventually migrate to the vapor space.

1. 4　Three – Phase Oil, Gas and Water Separation

Guidance to Reading

Three – phase separators work on the principle that the three components have different densities, which allows them to stratify when moving slowly with gas on top, water on the bottom and oil in the middle. Any solids such as sand will also settle in the bottom of the separator. Three – phase separators are designed as either horizontal or vertical pressure vessels. Because flow normally enters these vessels either directly from (1) a producing well or (2) a separator operating at a higher pressure, the vessel must be designed to separate the gas that **flashes** *from the liquid as well as separate the oil and water.*

Text

The separator design concepts that have been presented in the two – phase separation of liquid and gas **are applicable to** the three – phase separation. When oil and water are mixed with some **intensity** and then allowed to settle, a layer of relatively clean free water will appear at the bottom. The growth of this water layer with time will follow a **curve**. After a period of time, ranging anywhere from three minutes to thirty minutes, the change in the water height will be **negligible**. The water fraction, obtained from gravity settling, is called "free water." It is normally **beneficial** to separate the free water before attempting to treat the remaining oil and emulsion layers. "Three – phase separator" and "**free – water knockout**" are terms that are used to describe pressure vessels that are designed to separate and remove the **free water** from a mixture of crude oil and water.

Horizontal Separators

Three – phase separators are designed as either horizontal or vertical pressure vessels. Figure 1. 4. 1 is a schematic of a horizontal separator. The fluid enters the separator and hits an inlet diverter. This sudden change in momentum does the initial gross separation of liquid and vapor. In most designs, the inlet diverter contains a **downcomer** that directs the liquid flow below the oil/water interface.

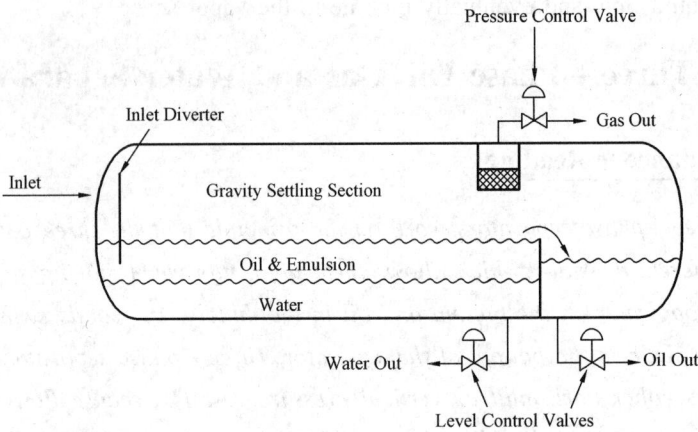

Figure 1. 4. 1   Horizontal three – phase separator schematic

The liquid collecting section of the vessel provides sufficient time so that the oil and emulsion form a layer or **"oil pad"** at the top. The free water settles to the bottom. Figure 1. 4. 1 illustrates a typical horizontal separator with an interface controller and weir. The weir maintains the oil level and the level controller maintains the water level. The oil is skimmed over the weir. The level of the oil downstream of the weir is controlled by a level controller that operates the oil dump valve.

The produced water flows from a **nozzle** in the vessel located upstream of the oil **weir**. An interface level controller senses the height of the oil/water interface. The controller sends a signal to the water dump valve thus allowing the correct amount of water to leave the vessel so that the oil/water interface is maintained at the design height.

The gas flows horizontally and out through a mist extractor to a pressure control valve that maintains constant vessel pressure. The level of the gas/oil interface can vary from half the diameter to 75% of the diameter depending on the relative importance of liquid/gas separation. The most common configuration is half full.

Figure 1.4.2 shows an **alternate configuration** known as a "**bucket and weir**" design. This design eliminates the need for a liquid interface controller. Both the oil and water flow over weirs where level control is accomplished by a simple **displacer float**. The oil overflows the oil weir into an **oil bucket** where its level is controlled by a level controller that operates the oil dump valve. The water flows under the oil bucket and then over a water weir. The **level downstream** of this weir is controlled by a level controller that operates the water dump valve.

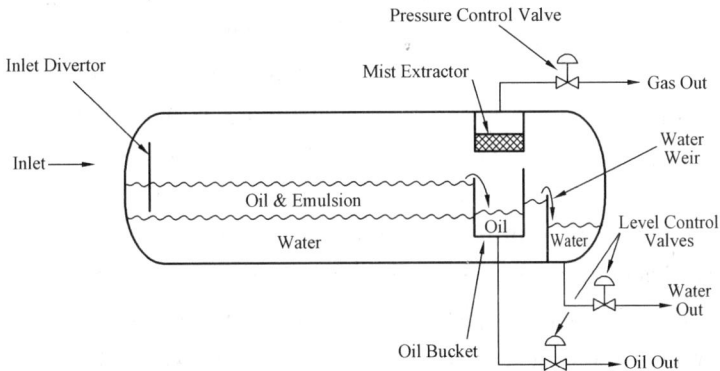

Figure 1.4.2　Bucket and weir design

The height of the oil weir controls the liquid level in the vessel. The difference in height of the oil and water weirs controls the thickness of the oil pad due to specific gravity differences. It is critical to the operation of the vessel that the water weir height be sufficiently below the oil weir height so that the oil pad thickness provides sufficient oil **retention time**. If the water weir is too low and the difference in specific gravity is not as great as **anticipated**, then the oil pad could grow in thickness to a point where oil will be swept under the oil box and out the water outlet. Normally, either the oil or the water weir is made adjustable so that changes in oil/water specific gravities or flow rates can be accommodated.

Interface control has the advantage of being easily **adjustable** to handle unexpected changes in oil or water specific gravity or **flow rates**. However, in heavy oil applications or where large amounts of emulsion or **paraffin** are antic-ipated, it may be difficult to sense interface level. In such a case bucket and weir control is recommended.

Vertical Separators

Figure 1.4.3 shows a typical configuration for a vertical three – phase separator. Flow enters the vessel through the side as in the horizontal separator, and the inlet diverter separates the bulk of the gas. A downcomer is required to transmit the liquid through the oil – gas interface so as not to disturb the **oil skimming action** taking place. A **chimney** is needed to equalize gas pressure between the lower section and the gas section.

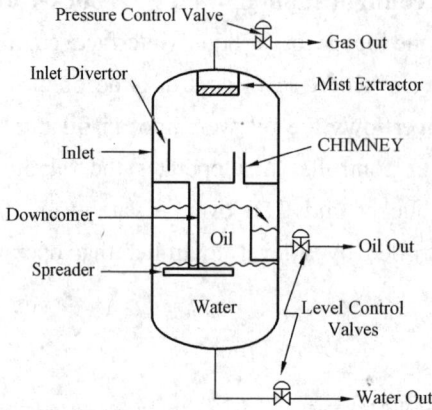

Figure 1.4.3　Vertical three – phase separator schematic

The **spreader** or **downcomer outlet** is located at the oil – water inter-face. From this point as the oil rises any free water trapped within the oil phase separates out. The water droplets flow countercurrent to the oil. Similarly, the water flows downward and oil droplets trapped in the water phase tend to rise countercurrent to the water flow.

Figure 1.4.4 shows the three different methods of control that are often used on vertical separators. The first is strictly level control. A regular displacer float is used to control the gas – oil interface and regulate a control valve dump-ing oil from the oil section. An interface float is used to control the oil – water interface and regulate a water outlet control valve. Because no internal **baffling** or weirs are used, this system is the easiest to fabricate and handles sand and solids production best.

The second method shown uses a weir to control the gas – oil interface lev-el at a constant position. This results in a better separation of water from the oil as all the oil must rise to the height of the oil weir before exiting the vessel. Its disadvantages are that the oil box takes up vessel volume and costs money to **fabricate**. In addition, sediment and solids could collect in the oil box and be difficult to **drain**, and a separate **low level shut – down** may be required to

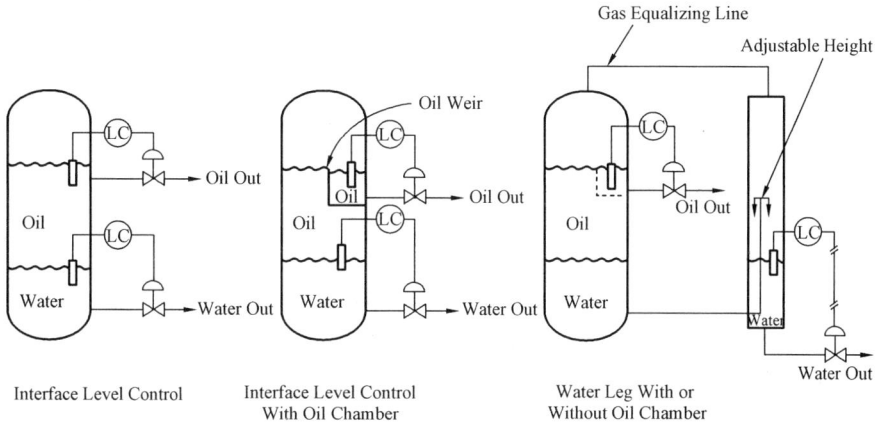

Figure 1.4.4 Liquid level control schemes

guard against the oil dump valve failing to open.

The third method uses two weirs, which eliminates the need for an interface float. Interface level is controlled by the height of the external water weir relative to the oil weir or outlet height. This is similar to the bucket and weir design of horizontal separators. The advantage of this system is that it eliminates the interface level control. The disadvantage is that it requires additional external piping and space.

Horizontal vs. Vertical Selection

The benefits of each type of design were described earlier. As in two – phase separation, it is also true for three – phase separation that the **flow geometry** in a horizontal vessel is more favorable from a **process** standpoint. However, there may be non – process reasons to select a vertical vessel for a specific application.

🔲 Words and Expressions

flash	闪蒸
intensity	强度
curve	曲线
negligible	微不足道的
beneficial	有益的
downcomer	降液管
nozzle	喷管

weir	挡油板
anticipate	预期
adjustable	可调节的
paraffin	石蜡
chimney	气体平衡管
spreader	油水分布器
baffling	折流板
fabricate	制作
drain	排出
process	工艺

🔲 Phrases and Expressions

be applicable to	适用于
free – water knockout	游离水脱除器
free water	游离水
oil pad	油层
alternate configuration	另一种形状
bucket and weir	油槽挡板式
displacer float	排液浮子
oil bucket	油槽
level downstream	下游液面
retention time	停留时间
flow rate	流量
oil skimming action	撇油
downcomer outlet	降液管下端出口
low level shut – down	低液面停机装置
flow geometry	流动几何形态

🔲 Language Focus

1. The separator design concepts that have been presented in the two – phase separation of liquid and gas are applicable to the three – phase separation.

（参考译文：气液两相分离的设计理念也适用于油气水三相分离。）

本句中"that"引导限制性定语从句，先行词是"concepts"，指代"the separator design concepts"；"be applicable to"是固定搭配，意为"适用于，应用

于"。

2. "Three – phase separator" and "free – water knockout" are terms that are used to describe pressure vessels that are designed to separate and remove the free water from a mixture of crude oil and water.

(参考译文:"三相分离器"及"游离水脱除器"两个术语指的是设计用来从原油和水的混合物中分离出游离水的压力容器。)

本句中有两个"that"引导的限制性定语从句,先行词分别是"terms"和"vessels",指代"terms"和"pressure vessels"。

3. The controller sends a signal to the water dump valve thus allowing the correct amount of water to leave the vessel so that the oil/water interface is maintained at the design height.

(参考译文:界面控制器将信号传送给排水阀,允许适量的水排出分离器,使油水界面维持在设计高度。)

本句中"thus"表示"因此";"so that"是固定搭配,表示"以至于"。

4. The level of the gas/oil interface can vary from half the diameter to 75% of the diameter depending on the relative importance of liquid/gas separation.

(参考译文:根据气液分离的具体要求,油气界面高度可控制在容器直径的1/2 至3/4 范围内。)

本句中"vary from. . . to. . . "是固定搭配,表示范围区间;"depending on the relative importance of liquid/gas separation"是伴随状语。

5. It is critical to the operation of the vessel that the water weir height be sufficiently below the oil weir height so that the oil pad thickness provides sufficient oil retention time.

(参考译文:分离器操作中十分关键的问题是挡水板高度必须充分低于隔油挡板高度,这样油层厚度才能为原油在分离器内提供充足的停留时间。)

本句中 "It is critical to. . . "意为"对……是至关重要的","that"从句中使用虚拟语气"be";"it"为形式主语。

Reinforced Learning

I . Answer the following questions for a comprehension of the text.

1. What should we consider while designing a vessel?

2. What are the respective advantages of interface control and bucket and

weir control in a horizontal separator?

3. What is the schematic of a "bucket and weir" design?

4. What is the advantage of interface control?

5. What are the three different methods of control often used on vertical separators?

II. Multiple choice:choose the correct one from the alternative answers to give the exact meaning of the words.

1. When oil and water are mixed with some underline{intensity} and then allowed to settle, a layer of relatively clean free water will appear at the bottom.

A. strength   B. length   C. quality   D. quantity

2. The growth of this water layer with time will follow a underline{curve}.

A. straight   B. line   C. wave   D. bend

3. After a period of time, ranging anywhere from three minutes to thirty minutes, the change in the water height will be underline{negligible}.

A. valuable   B. insignificant   C. obvious   D. stable

4. It is normally underline{beneficial} to separate the free water before attempting to treat the remaining oil and emulsion layers.

A. usual   B. profitable   C. favorable   D. marketable

5. Figure 1. 4. 2 shows an underline{alternate} configuration known as a "bucket and weir" design.

A. precise   B. better   C. exquisite   D. exchanging

6. This design underline{eliminates} the need for a liquid interface controller. Both the oil and water flow over weirs where level control is accomplished by a simple displacer float.

A. emphasizes   B. requests   C. excludes   D. proposes

7. Normally, either the oil or the water weir is made underline{adjustable} so that changes in oil/water specific gravities or flow rates can be accommodated.

A. flexible   B. adaptable   C. exploited   D. separable

8. Because no internal baffling or weirs are used, this system is the easiest to underline{fabricate} and handles sand and solids production best.

A. create   B. utilize   C. manufacture   D. operate

9. In addition, sediment and solids could collect in the oil box and be difficult to underline{drain}, and a separate low level shut – down may be required to guard against the oil dump valve failing to open.

A. melt   B. absorb   C. exhaust   D. mingle

10. As in two – phase separation, it is also true for three – phase separation that the flow geometry in a horizontal vessel is more favorable from a process standpoint.

A. art   B. stage   C. technology D. procedure

Ⅲ. Multiple choice: read the four suggested translations and choose the best answer.

1. When oil and water are mixed with some intensity and then allowed to settle, a layer of relatively clean free water will appear at the bottom.

A. 密度   B. 重量   C. 质量   D. 强度

2. After a period of time, ranging anywhere from three minutes to thirty minutes, the change in the water height will be negligible.

A. 微不足道的   B. 不值一提的

C. 显而易见的   D. 不可忽视的

3. The water fraction, obtained from gravity settling, is called "free water".

A. 免费水   B. 自由水   C. 游离水   D. 自来水

4. Because flow normally enters these vessels either directly from (1) a producing well or (2) a separator operating at a higher pressure, the vessel must be designed to separate the gas that flashes from the liquid as well as separate the oil and water.

A. 闪蒸   B. 闪光   C. 闪现   D. 闪动

5. The liquid collecting section of the vessel provides sufficient time so that the oil and emulsion form a layer or "oil pad" at the top.

A. 油顶   B. 油层   C. 油派   D. 油罐

6. The oil overflows the oil weir into an oil bucket where its level is controlled by a level controller that operates the oil dump valve.

A. 油桶   B. 油罐   C. 油槽   D. 油管

7. Normally, either the oil or the water weir is made adjustable so that changes in oil/water specific gravities or flow rates can be accommodated.

A. 可适应的 B. 可变化的 C. 可顺应的 D. 可调节的

8. However, in heavy oil applications or where large amounts of emulsion or paraffin are anticipated, it may be difficult to sense interface level. In such a case bucket and weir control is recommended.

A. 石油　　　　B. 石化　　　　C. 石蜡　　　　D. 沥青

9. A <u>chimney</u> is needed to equalize gas pressure between the lower section and the gas section.

A. 烟囱　　　　B. 烟管　　　　C. 气体平衡管　D. 气孔

10. As in two – phase separation, it is also true for three – phase separation that the flow geometry in a horizontal vessel is more favorable from a <u>process</u> standpoint.

A. 过程　　　　B. 技术　　　　C. 实验　　　　D. 工艺

Ⅳ. Put the following sentences into Chinese.

1. The separator design concepts that have been presented in the two – phase separation of liquid and gas are applicable to the three – phase separation.

2. "Three – phase separator" and "free – water knockout" are terms that are used to describe pressure vessels that are designed to separate and remove the free water from a mixture of crude oil and water.

3. The controller sends a signal to the water dump valve thus allowing the correct amount of water to leave the vessel so that the oil/water interface is maintained at the design height.

4. This results in a better separation of water from the oil as all the oil must rise to the height of the oil weir before exiting the vessel.

5. As in two – phase separation, it is also true for three – phase separation that the flow geometry in a horizontal vessel is more favorable from a process standpoint.

Ⅴ. Put the following paragraphs into Chinese.

1. When oil and water are mixed with some intensity and then allowed to settle, a layer of relatively clean free water will appear at the bottom. The growth of this water layer with time will follow a curve. After a period of time, ranging anywhere from three minutes to thirty minutes, the change in the water height will be negligible. The water fraction, obtained from gravity settling, is called "free water." It is normally beneficial to separate the free water before attempting to treat the remaining oil and emulsion layers.

2. The height of the oil weir controls the liquid level in the vessel. The difference in height of the oil and water weirs controls the thickness of the oil pad due to specific gravity differences. It is critical to the operation of the vessel

that the water weir height be sufficiently below the oil weir height so that the oil pad thickness provides sufficient oil retention time. If the water weir is too low and the difference in specific gravity is not as great as anticipated, then the oil pad could grow in thickness to a point where oil will be swept under the oil box and out the water outlet.

Chapter 2　Oil and Produced Water Treating

2. 1　Emulsion Treating

📖 Guidance to Reading

*Normal oil field emulsions consist of an oil continuous or external phase, and a water dispersed or internal phase. In some isolated cases, where there are high **water cuts**, it is possible to form **reverse emulsion** with water as the continuous phase and oil droplets the **internal phase**. Complex emulsions have been reported in low gravity, viscous crudes. These mixed emulsions contain a water **external phase** and have an internal water phase in the dispersed oil. The vast majority of oil treating systems deal with normal emulsions and that is what is discussed here.*

📖 Text

Forming Emulsions

For an emulsion to exist there must be two mutually **immiscible** liquids, an **emulsifying agent**, and sufficient agitation to disperse the discontinuous phase into the continuous phase. In oil production, oil and water are the two mutually immiscible liquids. An emulsifying agent in the form of small solid particles, paraffins, asphaltenes, etc. , is almost always present in the **formation fluid**s, and sufficient agitation always occurs as fluid **makes its way into** the well bore, up the **tubing**, and through the **surface choke**.

The properties and amount of emulsifying agent as well as the amount of agitation determine the "stability" of emulsions. Some stable emulsions may take weeks or months to separate if left alone in a tank with no treating. Other unstable emulsions may separate into relatively clean oil and water phases in just a matter of minutes.

Oil and water can form a water – in – oil emulsion, wherein water is the dispersed phase and oil is the external phase. On the contrary, they can form an oil – in – water emulsion, wherein the oil is the dispersed phase, and water is the **dispersion** medium.

Emulsifying Agent

When thinking about emulsion stability, it may be helpful to realize that in a pure oil and pure water mixture, without an emulsifying agent, no amount of agitation will create an emulsion. If the pure oil and water are mixed and placed in a container, they quickly separate. The natural state is for the immiscible liquids to establish the least contact or smallest surface area. The water dispersed in the oil forms spherical drops. Smaller drops will coalesce into larger drops and this will create a smaller interface area for a given volume. If no **emulsifier** is present, the droplets will eventually settle to the bottom causing the smallest interface area. This type of mixture is a true "dispersion".

An emulsifying agent has a surface active behavior. Some element in the emulsifier has a preference for the oil, and other elements are more attracted to the water. An emulsifier tends to be **insoluble** in one of the liquid phases. It thus concentrates at the interface. There are several ways an emulsifier changes a dispersion into an emulsion. The action of the emulsifier can be visualized as one or more of the following:

(1) It decreases the **interfacial tension** of the water droplet, thus causing smaller droplets to form. The smaller droplets take longer to coalesce into larger droplets, which can settle quickly.

(2) It forms a **viscous coating** on the droplets that keeps them from coalescing into larger droplets when they collide. Since **coalescence** is prevented, it takes longer for the small droplets created by agitation to settle out.

(3) The emulsifiers may be **polar molecules**, which **align** themselves in such a manner as to cause an electrical charge on the surface of the droplets. Since like electrical charges repel, two droplets must collide with sufficient force to overcome this **repulsion** before coalescence can occur.

Naturally – occurring **surface active materials** normally found in crude oil serve as emulsifiers. Paraffins, **resins**, **organic acids**, metallic salts, **colloidal silts** and clay, and **asphaltenes** are common emulsifiers in oil fields. Workover fluids and **drilling mud** are also sources of emulsifying agents.

The type and amount of emulsifying agent has an immediate effect on the emulsion's stability. It has been shown that the temperature history of the emulsion is also important as it effects the formation of paraffins and asphalt-enes. The speed of migration of the emulsifying agent to the oil/water interface and the behavior in terms of the **strength of the interface bond** are important factors. An emulsion treated soon after agitation or the creation of paraffins and **asphaltenes** can be less stable and easier to process if the migration of the emul-sifler is incomplete. An aged emulsion may become more difficult to treat. Normally, the lower the crude viscosity and lighter the crude the more rap-id the aging process. Therefore, early treatment may be a lesser factor in treating low – viscosity, high – API – gravity crudes.

Demulsifiers

Chemical demulsifiers sold under various trade names, such as Tretolite^{TM}, Visco^{TM}, and Breaxit^{TM}, are highly useful in **resolving** emulsions. Demulsifiers act to neutralize the effect of emulsifying agents. Typically, they are surface ac-tive agents and thus their excessive use can decrease the surface tension of water droplets and actually create more stable emulsions.

There are four important actions required of a demulsifier:

(1) Strong attraction to the oil – water interface.

(2) Flocculation.

(3) Coalescence.

(4) **Solid wetting**.

When these actions are present, they promote the separation of oil and wa-ter. The demulsifier must have the ability to migrate rapidly through the oil phase to the droplet interface, where it must compete with the more concentrated emulsifying agent. The **demulsifier** must also have an attraction for droplets with a similar condition. In this way large clusters of droplets gather which, un-der a microscope, appear like bunches of **fish eggs**. The oil will take on a bright appearance since small droplets are no longer present to **scatter** the light rays. At this point the emulsifier film is still continuous. If the emulsifier is weak, the **flocculation force** may be enough to cause coalescence. This is not true in most cases and the demulsifier must therefore neutralize the emulsifier and promote a **rupture** of the droplet interface film. This is the opener that cau-ses coalescence. With the emulsion in a flocculated condition the film rupture re-sults in rapid growth of water – drop size.

The manner in which the demulsifier **neutralizes** the emulsifier depends upon the type of emulsifiers. **Iron sulfides**, clays, and drilling muds can be **water wet** causing them to leave the interface and be diffused into the water droplet. Paraffins and asphaltenes could be dissolved or altered to make their films less viscous so they will flow out of the way on collision or could be made **oil wet** so they will be dispersed in the oil.

It would be unusual if one chemical structure could produce all four desirable actions. A blend of compounds is therefore used to achieve the right balance of activity.

The demulsifier selection should be made with the process system in mind. If the treating process is a settling tank, a relatively **slow – acting compound** can be applied with good results. On the other hand, if the system is a **chemelectric process** where some of the **flocculation** and coalescing action is accomplished by an electric field, there is need for a quick – acting compound, but not one that must complete the **dropletbuilding action**.

As field conditions change, the chemical requirements can change. If the process is modified, e. g. , very low **rate**s on electrostatic units, the chemical requirement can change. Seasonal changes bring paraffin **induced** emulsion problems. **Workover**s contribute to solids content, which alters emulsion stability. So, no matter how satisfactory a demulsifier is at one point in time, it cannot be assumed that it will always be satisfactory over the life of the field.

🔲 **Words and Expressions**

immiscible	不混相的
agitation	搅动
paraffin	石蜡
asphaltene	沥青质
tubing	油管
coalescence	聚结
dispersion	分散体
emulsifier	乳化剂
insoluble	不溶于
coalescence	聚结
align	排列
repulsion	斥力

resin	胶质
asphaltene	沥青质
demulsifier	破乳剂
resolve	处理
scatter	分散
rupture	破裂
neutralize	抵消
flocculation	絮凝
rate	流量
induced	诱发的
workover	修井作业

▯ Phrases and Expressions

water cut	含水量
reverse emulsion	反相乳状液
external phase	外相
internal phase	内相
emulsifying agent	乳化剂
formation fluid	地层流体
make its way into	流向
surface choke	地面油嘴
interfacial tension	界面张力
viscous coating	黏性包膜
polar molecule	极性分子
surface active material	表面活性物质
organic acid	有机酸
colloidal silt	胶态粉砂
drilling mud	钻井液
strength of the interface bond	界面结合力强度
solid wetting	固体润湿
fish egg	鱼子
flocculation force	絮凝力
iron sulfide	硫化铁
water wet	亲水的

oil wet	亲油的
slow – acting compound	长效药剂
chemelectric process	电化学处理工艺
dropletbuilding action	液滴增大作用

Language Focus

1. Some stable emulsions may take weeks or months to separate if left alone in a tank with no treating.

（参考译文：某些稳定的乳状液如果存放在罐里不做任何处理，可能需要几周乃至几个月才能分层。）

本句中"if"引导条件状语从句"left alone in a tank with no treating"，"if"后省略了"stable emulsions are"。

2. In some isolated cases, where there are high water cuts, it is possible to form reverse emulsion with water as the continuous phase and oil droplets the internal phase.

（参考译文：在一些高含水的情况下，有可能形成水为连续相而油滴为内相的反相乳状液。）

本句中"where"引导非限制性定语从句"there are high water cuts"，指代前面的"in some isolated cases"；"as"作介词并与"form"搭配，"oil droplets"后省略了"as"。

3. On the other hand, if the system is a chemelectric process where some of the flocculation and coalescing action is accomplished by an electric field, there is need for a quick – acting compound, but not one that must complete the dropletbuilding action.

（参考译文：另一方面，假如系统采用电化学处理工艺，靠电场作用完成部分絮凝和聚结过程，就需要选用速效剂而不是具有液滴增大作用的药剂。）

本句中副词短语"on the other hand"意为"另一方面"。这是一个复合句，主句是"there is need for a quick – acting compound"；"if"引导条件状语从句"the system is a chemelectric process where some of the flocculation and coalescing action is accomplished by an electric field"，其中"where"引导限制性定语从句"some of the flocculation and coalescing action is accomplished by an electric field"，指代"in a chemelectric process"；"but"作连词，"one"指代"compound"，"that"引导限制性定语从句"must complete the dropletbuilding action"，指代"one"。

⌗ Reinforced Learning

I . **Answer the following questions for a comprehension of the text.**

1. What are normal emulsions?

2. What is the action of the emulsifier?

3. What are sources of emulsifying agents?

4. What are four important actions required of a demulsifier?

5. How to select the demulsifier?

II . **Multiple choice:choose the correct one from the alternative answers to give the exact meaning of the words.**

1. For an emulsion to exist there must be two mutually immiscible liquids, an emulsifying agent,and sufficient <u>agitation</u> to disperse the discontinuous phase into the continuous phase.

    A. tension   B. unrest   C. suspense   D. stir

2. Complex emulsions have been reported in low gravity,<u>viscous</u> crudes.

    A. sticky   B. tricky   C. muggy   D. precarious

3. The <u>action</u> of the emulsifier can be visualized as one or more of the following.

    A. act   B. deed   C. role   D. performance

4. Since like electrical charges <u>repel</u>,two droplets must collide with sufficient force to overcome this repulsion before coalescence can occur.

    A. attract   B. repulse   C. appeal   D. react

5. Normally,the lower the crude viscosity and lighter the crude the more rapid the <u>aging</u> process.

    A. ancient   B. obsolete   C. mature   D. modern

6. Chemical demulsifiers sold under various trade names,such as Tretolite^{TM},Visco^{TM},and Breaxit^{TM},are highly useful in <u>resolving</u> emulsions.

    A. deciding   B. blending   C. mixing   D. settling

7. Demulsifiers act to <u>neutralize</u> the effect of emulsifying agents.

    A. act   B. ignite   C. counteract   D. agree

8. The oil will take on a bright appearance since small droplets are no longer present to <u>scatter</u> the light rays.

    A. amass   B. disperse   C. collect   D. accumulate

9. This is not true in most cases and the demulsifier must therefore neutralize the emulsifier and promote a <u>rupture</u> of the droplet interface film.

　　A. separation　　　B. merging　　　C. connection　　　D. liaison

10. If the process is modified, e. g. , very low rates on electrostatic units, the chemical requirement can change. Seasonal changes bring paraffin <u>induced</u> emulsion problems.

　　A. deduced　　　B. accumulated　　C. amazed　　　D. evoked

Ⅲ. Multiple choice：read the four suggested translations and choose the best answer.

1. An emulsifying agent in the form of small solid particles, paraffins, asphaltenes, etc. , is almost always present in the <u>formation fluids</u>, and sufficient agitation always occurs as fluid makes its way into the well bore, up the tubing, and through the surface choke.

　　A. 形成液体　　B. 构成液体　　C. 地层流体　　　D. 构造流体

2. An emulsifying agent in the form of small solid particles, paraffins, asphaltenes, etc. , is almost always present in the formation fluids, and sufficient agitation always occurs as fluid <u>makes its way</u> into the well bore, up the tubing, and through the surface choke.

　　A. 通过　　　B. 流向　　　C. 经过　　　D. 走向

3. Normal oil field emulsions consist of an oil continuous or <u>external phase</u>, and a water dispersed or internal phase.

　　A. 外部　　　B. 内部　　　C. 内相　　　D. 外相

4. In some isolated cases, where there are high water cuts, it is possible to form reverse emulsion with water as the continuous phase and oil droplets the internal phase. Complex emulsions have been reported in <u>low gravity, viscous crudes</u>.

　　A. 低硫原油　　　　　　　B. 高硫原油
　　C. 轻质原油　　　　　　　D. 重质黏稠原油

5. If no emulsifier is present, the droplets will eventually settle to the bottom causing the smallest interface area. This type of mixture is a true "<u>dispersion</u>".

　　A. 分离式　　　B. 混合体　　C. 分散体　　　D. 散布体

6. An emulsifier tends to be <u>insoluble</u> in one of the liquid phases.

　　A. 不溶于　　　B. 不能解决的　　C. 难解释的　　　D. 难解决的

7. The emulsifiers may be polar molecules, which <u>align</u> themselves in such a manner as to cause an electrical charge on the surface of the droplets.

    A. 排行   B. 匹配   C. 排列   D. 配对

8. It has been shown that the <u>temperature history</u> of the emulsion is also important as it effects the formation of paraffins and asphaltenes.

    A. 气温历史   B. 气候历史

    C. 温度记录   D. 温度变化过程

9. Demulsifiers act to <u>neutralize</u> the effect of emulsifying agents.

    A. 中立   B. 抵消   C. 中和   D. 无效

10. Iron sulfides, clays, and drilling muds can be <u>water wet</u> causing them to leave the interface and be diffused into the water droplet.

    A. 水湿的   B. 加水的   C. 浇水的   D. 亲水的

IV. Put the following sentences into Chinese.

1. For an emulsion to exist there must be two mutually immiscible liquids, an emulsifying agent, and sufficient agitation to disperse the discontinuous phase into the continuous phase.

2. The emulsifiers may be polar molecules, which align themselves in such a manner as to cause an electrical charge on the surface of the droplets.

3. Typically, they are surface active agents and thus their excessive use can decrease the surface tension of water droplets and actually create more stable emulsions.

4. Paraffins and asphaltenes could be dissolved or altered to make their films less viscous so they will flow out of the way on collision or could be made oil wet so they will be dispersed in the oil.

5. On the other hand, if the system is a chemelectric process where some of the flocculation and coalescing action is accomplished by an electric field, there is need for a quick – acting compound, but not one that must complete the dropletbuilding action.

V. Put the following paragraphs into Chinese.

1. The manner in which the demulsifier neutralizes the emulsifier depends upon the type of emulsifiers. Iron sulfides, clays, and drilling muds can be water wet causing them to leave the interface and be diffused into the water droplet. Paraffins and asphaltenes could be dissolved or altered to make their films

less viscous so they will flow out of the way on collision or could be made oil wet so they will be dispersed in the oil.

2. The type and amount of emulsifying agent has an immediate effect on the emulsion's stability. It has been shown that the temperature history of the e-mulsion is also important as it effects the formation of paraffins and asphalt-enes. The speed of migration of the emulsifying agent to the oil/water interface and the behavior in terms of the strength of the interface bond are important factors. An emulsion treated soon after agitation or the creation of paraffins and asphaltenes can be less stable and easier to process if the migration of the emulsi-fler is incomplete. An aged emulsion may become more difficult to treat. Normally, the lower the crude viscosity and lighter the crude the more rapid the aging process. Therefore, early treatment may be a lesser factor in treating low − viscosity, high − API − gravity crudes.

2. 2　Crude Oil Treating Systems

🔲 Guidance to Reading

Removing water from crude oil often requires additional processing beyond gravitational separation. A common method for separating the "water − in − oil" emulsion is to heat the stream. The process of coalescence requires that the water droplets have adequate time to contact each other. It also assumes that the buoyant forces on the coalesced droplets are sufficient to enable these droplets to settle to the bottom of the treating vessel. Consequently, design considerations should necessarily include temperature, time, viscous properties of oil that inhibit settling, and the physical dimensions of the vessel, which determine the velocity at which settling must occur.

🔲 Text

In selecting a treating system, several factors should be considered to determine the most desirable methods of treating the crude oil to contract requirements. Some of these factors are:

(1) **Tightness of the emulsion.**

(2) **Specific gravity** of the oil and produced water.

(3) Corrosiveness of the crude oil, produced water, and casinghead gas.

(4) **Scaling tendencies** of the produced water.

(5) Quantity of fluid to be treated and percent water in the fluid.

(6) **Paraffin – forming** tendencies of the crude oil.

(7) Desirable operating pressures for equipment.

(8) Availability of a sales outlet and value of the casinghead gas produced.

A common method for separating this " water – in – oil" emulsion is to heat the stream. Increasing the temperature of the two immiscible liquids **deactivates** the **emulsifying agent**, allowing the dispersed water droplets to **collide**. As the droplets collide they grow in size and begin to settle. If designed properly, the water will settle to the bottom of the treating vessel due to differences in specific gravity.

Treating Equipment

Vertical Treaters

The most commonly used single – well lease treater is the vertical treater as shown in Figure 2.2.1. Flow enters the top of the treater into a gas separation section. Care must be exercised to size this section so that it has adequate dimensions to separate the gas from the inlet flow. If the treater is located downstream of a separator, this chamber can be very small. The gas separation section should have an **inlet diverter** and a **mist extractor**.

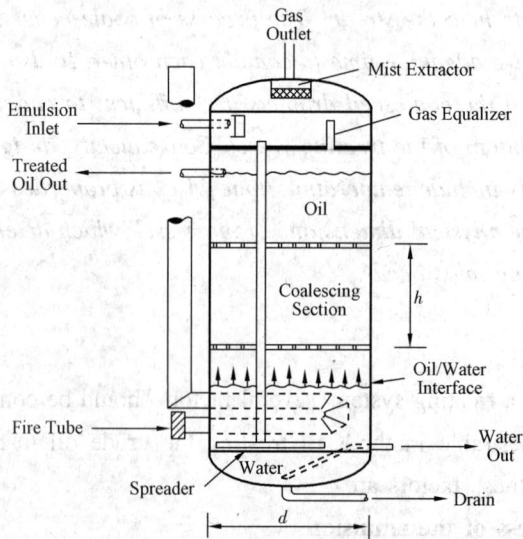

Figure 2.2.1　Vertical treater schematic

　　The liquids flow through a downcomer to the base of the treater, which. serves as a free – water **knockout** section. If the treater is located downstream of a free – water knockout, the bottom section can be very small. If the total well-stream is to be treated this section should be sized for 3 to 5 minutes retention time for both the oil and the water to allow the free water to settle out. This will minimize the amount of fuel gas needed to heat the liquid stream rising through the heating section. The end of the downcomer should be slightly below the oil water interface to "water wash" the oil being treated. This will assist in the coalescence of water droplets in the oil.

　　The oil and emulsion rises over the heater fire – tubes to a coalescing section where sufficient retention time is provided to allow the small water particles in the oil continuous phase to coalesce and settle to the bottom.

　　Treated oil flows out the oil outlet. Any gas, flashed from the oil due to heating, flows through the **equalizing line** to the gas space above. Oil level is maintained by **pneumatic** or **lever operated dump valves**. Oil – water interface is controlled by an interface controller, or an adjustable external water leg.

　　The detailed design of the treater, including the design of internals (many features of which are patented) should be the responsibility of the equipment supplier. Figure 2. 2. 2 shows a "gunbarrel" tank, which is a vertical flow treater in an **atmospheric tank**. Typically, gunbarrels have a gas separating chamber or "boot" on top where gas is separated and **vent**ed, and a downcomer. Because gunbarrels tend to be of larger diameter than vertical heater – treaters, many have elaborate **spreader system**s to try and create uniform (i. e. , plug) upward flow of the emulsion to **take maximum advantage of** the entire **cross section**. Most gunbarrels are unheated, though it is possible to provide heat by heating the incoming stream external to the tank, installing **heating coil**s in the tank, or circulating the water to an external heater in a **closed loop**. It is preferable to heat the inlet so that more gas is liberated in the boot, although this means that fuel will be used in heating any free water in the inlet.

　　Gunbarrels are most often used in older, small flow rate, **onshore facilities**. In recent times vertical heater – treaters have become so inexpensive that they have replaced gunbarrels in single well installations. On larger installations onshore in warm weather areas gun barrels are still commonly used. In areas that

Figure 2.2.2 Typical gunbarrel settling tank with internal flume

have a winter season it tends to be too expensive to keep the large volume of oil at a high enough temperature to combat potential **pour point** problems.

Horizontal Treaters

For most multi – well situations horizontal treaters are normally required. Figure 2.2.3 shows a typical design of a horizontal treater.

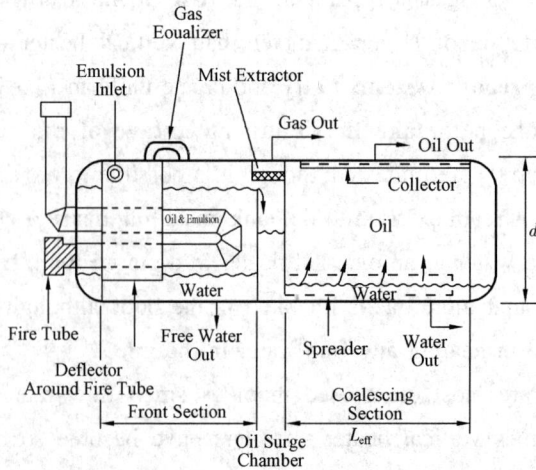

Figure 2.2.3 Horizontal heater – treater schematic

Flow enters the front section of the treater where gas is flashed. The liquid falls around the outside to the **vicinity** of the oil – water interface where the liquid is "water washed" and the free water is separated. Oil and emulsion rise past the fire tubes and are skimmed into the oil surge chamber. The oil – water interface in the inlet section of the vessel is controlled by an **interface level** controller, which operates a **dump valve** for the free water.

The oil and emulsion flow through a spreader into the back or coalescing section of the vessel, which is fluid packed. The spreader distributes the flow evenly throughout the length of this section. Treated oil is collected at the top through a collection device sized to maintain uniform vertical flow of the oil. Coalescing water droplets fall countercurrent to the rising oil continuous phase. The oil – water interface is maintained by a **level controller** and dump valve for this section of the vessel.

A level control in the oil surge chamber operates a dump valve on the oil outlet line regulating the flow of oil out the top of the vessel to maintain a **fluid packed** condition.

The inlet section must be sized to handle settling of the free water and heating of the oil. The coalescing section must be sized to provide adequate retention time for coalescence to occur and to allow the coalescing water droplets to **settle downward countercurrent** to the upward flow of the oil.

Electrostatic Treaters

Some treaters use an electrode section. Figure 2. 2. 4 illustrates a typical design of a horizontal electrostatic treater. The flow path in an electrostatic treater is the same as a horizontal treater. The only difference is that an **AC and/or DC electrostatic field** is used to promote coalescence of the water droplets.

Procedures for designing electrostatic coalescers have not been published. Since coalescence of droplets in an electric field is so dependent on the characteristics of the particular emulsion to be treated, it is unlikely that a general relationship of water droplet size to use in the **settling equations** can be developed. Field experience tends to indicate that electrostatic treaters are efficient at reducing water content in the crude below the 0. 5 to 1. 0% basic sediment and water (BS&W) level. This makes them particularly attractive for desalting applications. However, for normal crude treating, where 0. 5 to 1. 0% BS&W is acceptable, it is recommended that they be sized as heater – treaters. By **trial and**

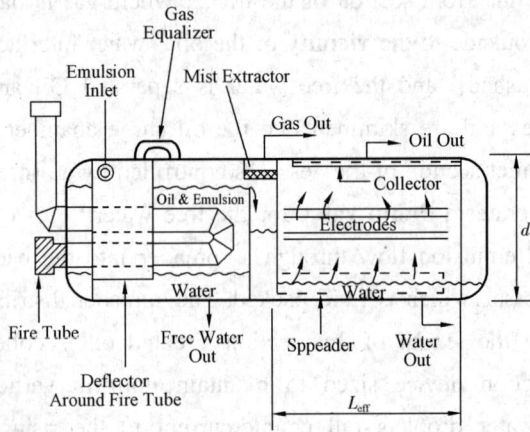

Figure 2.2.4   Horizontal electrostatic treater schematic

error after installation, the **electric grid**s may be able to allow treating to occur at lower temperatures.

🔲 Words and Expressions

inhibit	抑制
deactivate	削弱
collide	碰撞
knockout	打击;脱模;出坯;分离
pneumatic	气动
vent	放空
vicinity	附近

🔲 Phrases and Expressions

buoyant force	浮力
physical dimension	实际尺寸
tightness of the emulsion	乳状液稳定度
specific gravity	相对密度
scaling tendency	结垢趋势
paraffin – forming	结蜡
emulsifying agent	乳化剂
inlet diverter	入口分流器
mist extractor	除雾器

equalizing line	气体平衡管
lever operated	杠杆操纵的
dump valve	排泄阀
atmospheric tank	常压罐
spreader system	油水分流系统
take maximum advantage of	充分利用
cross section	横截面积
heating coil	加热盘管
closed loop	闭路
onshore facilities	陆上处理中心
pour point	倾点
interface level	界面
dump valve	泄放阀
level controller	液面控制器
fluid packed	液封的
settle downward countercurrent	逆向沉降
electrostatic treater	静电处理器
AC and/or DC electrostatic field	交流或直流静电场
settling equation	沉降方程
trial and error	调试
electric grid	电极板

Language Focus

1. Consequently, design considerations should necessarily include temperature, time, viscous properties of oil that inhibit settling, and the physical dimensions of the vessel, which determine the velocity at which settling must occur.

（参考译文：因此，设计时有必要考虑温度、时间、抑制水滴沉降的原油黏滞特性以及决定水滴沉降速度的容器实际尺寸等。）

本句中"consequently"是副词；"that"引导限制性定语从句，指代其先行词"viscous properties of oil"；"which"引导非限制性定语从句，指代其先行词"the physical dimensions of the vessel"；"at which"引导限制性定语从句，先行词是"velocity"，指代"at the velocity"。

2. If designed properly, the water will settle to the bottom of the treating vessel due to differences in specific gravity.

（参考译文：如果设计正确，那么，因为存在油水相对密度差，水将沉到处理器底部。）

本句中"If designed properly"是条件状语，"if"后省略了"crude oil treating systems are"；"due to"是介词短语，意为"因为、由于"。

3. Most gunbarrels are unheated, though it is possible to provide heat by heating the incoming stream external to the tank, installing heating coils in the tank, or circulating the water to an external heater in a closed loop.

（参考译文：大多数油水分离罐是不加热的，不过也可以在罐外预热进罐流体，或在罐内加装加热盘管，或设计一个加热回路使热水闭路循环。）

本句中"though"引导让步状语从句"it is possible to provide heat by heating the incoming stream external to the tank, installing heating coils in the tank, or circulating the water to an external heater in a closed loop"；"by"是介词，表示"通过……方法"，后面接"heating, installing"和"circulating"三个并列的动名词。

中 Reinforced Learning

I. Answer the following questions for a comprehension of the text.

1. What are the factors that should be considered to determine the most desirable methods of treating the crude oil to contract requirements in selecting a treating system?

2. What are design considerations for coalescence?

3. What is common method for separating the "water – in – oil" emulsion? Why?

4. When is the vertical treater most commonly used? And when is the horizontal treater most commonly used?

5. What is the difference between a horizontal treater and an electrostatic treater?

II. Multiple choice: choose the correct one from the alternative answers to give the exact meaning of the words.

1. Increasing the temperature of the two immiscible liquids <u>deactivates</u> the emulsifying agent, allowing the dispersed water droplets to collide.

    A. activates   B. weakens   C. strengthens   D. stimulates

2. Increasing the temperature of the two immiscible liquids deactivates the

emulsifying agent, allowing the dispersed water droplets to collide.

    A. crash   B. cross   C. happen   D. tear

    3. It also assumes that the buoyant forces on the coalesced droplets are sufficient to enable these droplets to settle to the bottom of the treating vessel.

    A. physical   B. light   C. sinking   D. elastic

    4. Consequently, design considerations should necessarily include temperature, time, viscous properties of oil that inhibit settling, and the physical dimensions of the vessel, which determine the velocity at which settling must occur.

    A. inhabit   B. proliferate   C. prevent   D. activate

    5. If the total wellstream is to be treated this section should be sized for 3 to 5 minutes retention time for both the oil and the water to allow the free water to settle out.

    A. remain   B. recovery   C. suspect   D. suspension

    6. The oil and emulsion rises over the heater fire – tubes to a coalescing section where sufficient retention time is provided to allow the small water particles in the oil continuous phase to coalesce and settle to the bottom.

    A. successive   B. following   C. ancient   D. former

    7. Any gas, flashed from the oil due to heating, flows through the equalizing line to the gas space above.

    A. warmed   B. blazed   C. flied   D. steamed

    8. Oil – water interface is controlled by an interface controller, or an adjustable external water leg.

    A. foot   B. angle   C. barrel   D. tube

    9. Typically, gunbarrels have a gas separating chamber or "boot" on top where gas is separated and vented, and a downcomer.

    A. blanked   B. filled   C. emptied   D. sorted

    10. Coalescing water droplets fall countercurrent to the rising oil continuous phase.

    A. counterclockwise   B. reversible

    C. counterattack   D. irreversible

    III. Multiple choice: read the four suggested translations and choose the best answer.

    1. Some of these factors are: Tightness of the emulsion, Specific gravity of the oil and produced water.

    A. 特别重力   B. 特殊重力   C. 相对密度   D. 相对压力

    2. Increasing the temperature of the two <u>immiscible</u> liquids deactivates the emulsifying agent, allowing the dispersed water droplets to collide.

    A. 不溶   B. 不相交   C. 不混相   D. 不分离

    3. The process of <u>coalescence</u> requires that the water droplets have adequate time to contact each other.

    A. 聚结   B. 聚集   C. 离合   D. 集散

    4. The liquids flow through a <u>downcomer</u> to the base of the treater, which serves as a free – water knockout section.

    A. 下降者   B. 降液管   C. 下水道   D. 插水管

    5. The oil and emulsion rises over the heater fire – tubes to a coalescing section where sufficient retention time is provided to allow the small water particles in the oil continuous <u>phase</u> to coalesce and settle to the bottom.

    A. 段   B. 相   C. 期   D. 阶

    6. Any gas, <u>flashed</u> from the oil due to heating, flows through the equalizing line to the gas space above.

    A. 闪光   B. 闪现   C. 闪动   D. 闪蒸

    7. Oil level is maintained by <u>pneumatic</u> or lever operated dump valves.

    A. 气动的   B. 机动的   C. 机械的   D. 机控的

    8. Because gunbarrels tend to be of larger diameter than vertical heater – treaters, many have elaborate spreader systems to try and create uniform (i. e. , plug) upward flow of the emulsion to <u>take maximum advantage of</u> the entire cross section.

    A. 得到益处   B. 发挥优势   C. 充分利用   D. 全面运用

    9. The coalescing section must be sized to provide adequate retention time for coalescence to occur and to allow the coalescing water droplets to <u>settle downward countercurrent</u> to the upward flow of the oil.

    A. 向下逆行   B. 逆向沉降   C. 降下反向   D. 下降相反

    10. By <u>trial and error</u> after installation, the electric grids may be able to allow treating to occur at lower temperatures.

    A. 实验   B. 试行   C. 检查   D. 调试

IV. Put the following sentences into Chinese.

    1. In selecting a treating system, several factors should be considered to determine the most desirable methods of treating the crude oil to contract require-

ments.

2. If designed properly, the water will settle to the bottom of the treating vessel due to differences in specific gravity.

3. The end of the downcomer should be slightly below the oil water interface to "water wash" the oil being treated.

4. In areas that have a winter season it tends to be too expensive to keep the large volume of oil at a high enough temperature to combat potential pour point problems.

5. The coalescing section must be sized to provide adequate retention time for coalescence to occur and to allow the coalescing water droplets to settle downward countercurrent to the upward flow of the oil.

V. Put the following paragraphs into Chinese.

1. Because gunbarrels tend to be of larger diameter than vertical heater – treaters, many have elaborate spreader systems to try and create uniform upward flow of the emulsion to take maximum advantage of the entire cross section. Most gunbarrels are unheated, though it is possible to provide heat by heating the incoming stream external to the tank, installing heating coils in the tank, or circulating the water to an external heater in a closed loop. It is preferable to heat the inlet so that more gas is liberated in the boot, although this means that fuel will be used in heating any free water in the inlet.

2. Since coalescence of droplets in an electric field is so dependent on the characteristics of the particular emulsion to be treated, it is unlikely that a general relationship of water droplet size to use in the settling equations can be developed. Field experience tends to indicate that electrostatic treaters are efficient at reducing water content in the crude below the 0. 5 to 1. 0% basic sediment and water level. This makes them particularly attractive for desalting applications. However, for normal crude treating, where 0. 5 to 1. 0% BS&W is acceptable, it is recommended that they be sized as heater – treaters.

2. 3　Produced – Water Treating Systems(1)

中 Guidance to Reading

Environmental impacts caused by the disposal of produced water have been reported since the mid – 1800s when the first oil and gas wells were drilled and operated. Many produced waters contain elevated levels of dissolved ions (salts) , hydrocarbons , and trace elements which are phytotoxic and occur in elevated concentrations to be adsorbed in the soil. Untreated produced water discharges may be harmful to the surrounding environment , causing degradation of soils , ground water , surface water , and ecosystems they support. The purpose of this article is to present a procedure for selecting the appropriate type of equipment for treating oil from produced water to make treated produced water to be to be environmentally acceptable.

🔲 Text

In producing operations it is often necessary to handle wastewater that may include water produced with crude oil, rain water, and **washdown water**. The water must be separated from the crude oil and disposed of in a manner that does not violate established environmental regulations. In offshore areas where discharge to the sea is allowed, the governing regulatory body specifies the maximum hydrocarbon content in the water that may be discharged overboard. The range is currently 15 mg/l to 50 mg/l depending on the specific location. In most onshore locations the water cannot be disposed of on the surface, due to possible salt **contamination**, and must be injected into an acceptable disposal **formation** or disposed of by evaporation. In either case it will probably be necessary to treat the produced water to lower its hydrocarbon content below that normally obtained from free – water knockouts and oil treaters.

System Description

Produced water will always have some form of primary treating prior to disposal. This could be a **skim tank**, skim vessel, **CPI**, or **crossflow separator**. All of these devices employ gravity separation techniques. Depending upon the **severity** of the treating problem, secondary treating utilizing a CPI, crossflow separator, or a flotation unit may be required.

Offshore, produced water can be piped directly overboard after treating, or it can be routed through a **disposal pile** or a **skim pile**. **Deck drains** must be treated for removal of "free" oil. Onshore, the water will normally be reinjected in the formation or be pumped into a disposal well.

For safety considerations, closed drains, if they exist in the process, should never be tied into atmospheric drains and should be routed to a pressure vessel

prior to entering an **atmospheric tank** or pile. This could be done in a skim vessel, crossflow separator or CPI in a pressure vessel. The latter two could be used
where it is desirable to separate sand from the system.

Treating Equipment

Settling Tanks and Skimmer Vessels

The simplest form of primary treating equipment is a settling (skim) tank
or vessel. These items are normally designed to provide long residence times
during which coalescence and gravity separation can occur. Skimmers can be either vertical or horizontal in configuration. In vertical skimmers the oil droplets
must rise upward countercurrent to the downward flow of the water. Some vertical skimmers have **inlet spreaders** and **outlet collector**s to help even the distribution of the flow, as shown in Figure 2.3.1 The inlet directs the flow below
the oil – water interface. Small amounts of gas liberated from the water help to
"float" the oil droplets. In the **quiet zone** between the spreader and the water
collector, some coalescence can occur and the buoyancy of the oil droplets causes them to rise counter to the water flow. Oil will be collected and skimmed off
the surface.

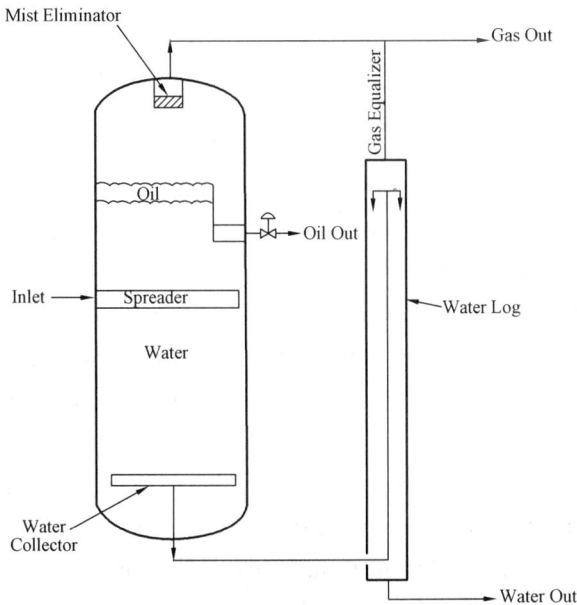

Figure 2.3.1　Vertical skimmer schematic

The thickness of the oil pad depends on the relative heights of the oil weir and the water leg, and the difference in specific gravity of the two liquids. Often, an interface lever controller is used in place of the water leg.

In horizontal skimmers the oil droplets rise **perpendicular to** the flow of the water, as shown in Figure 2.3.2. The inlet enters below the oil pad. The water then turns and flows horizontally for most of the length of the vessel. **Baffles** could be installed to straighten the flow. Oil droplets coalesce in this section of the vessel and rise to the oil – water surface where they are captured and eventually skimmed over the oil weir. The height of the oil can be controlled by interface control, by a water leg similar to that shown in Figure 2.3.1, or by a bucket and weir arrangement.

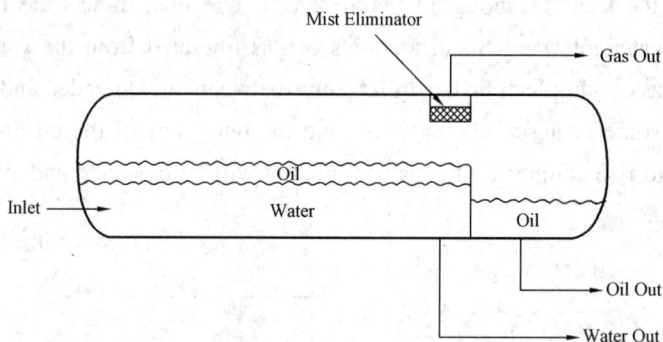

Figure 2.3.2   Horizontal skimmer schematic

Horizontal vessels are more efficient at water treating because the oil droplets do not have to flow countercurrent to the water flow. However, vertical skimmers are used in instances where:

(1) Sand and other solid particles must be handled. This can be done in vertical vessels with either the water outlet or a **sand drain** off the bottom. Experience with elaborately designed sand drains in large horizontal vessels has not been very satisfactory.

(2) Liquid surges are expected. Vertical vessels are less **susceptible** to high level shutdowns due to liquid surges. Internal waves due to surging in horizontal vessels can trigger a level float even though the volume of liquid between the normal operating level and the high level shutdown is equal to or larger than that in a vertical vessel.

The choice of pressure versus atmospheric vessel for the skimmer tank is not determined solely by the water treating requirements. The overall needs of the system need to be considered in this decision. Pressure vessels are more expensive. However, they are recommended where:

(1) Potential gas **blowby** through the upstream vessel dump system could create too much **back - pressure** in an **atmospheric vent system**.

(2) The water must be dumped to a higher level for further treating and a pump would be needed if an atmospheric vessel were installed.

Due to the potential danger from overpressure and potential gas venting problems associated with atmospheric vessels, pressure vessels are preferred. However, an individual cost/benefit decision must be made.

A minimum residence time of 10 to 30 minutes should be provided to assure that surges do not upset the system and to provide for some coalescence. As previously discussed, the potential benefits of providing much more residence time will probably not be cost efficient beyond this point. Skimmers with long residence times require baffles to attempt to distribute the flow and eliminate **short circuiting**. Tracer studies have shown that skimmer tanks, even those with carefully designed spreaders and baffles, exhibit poor flow behavior and short circuiting. This is probably due to density and temperature differences, deposition of solids, corrosion of spreaders, etc.

Words and Expressions

contamination	污染
formation	地层
CPI	波纹板隔油池
severity	难度
baffle	折流板
susceptible	易于
blowby	漏气

Phrases and Expressions

washdown water	冲洗水
skim tank	隔油罐
crossflow separator	交叉流分离器
disposal pile	排放管桩

skim pile	隔油管桩
deck drain	甲板污水
atmospheric tank	常压罐
inlet spreader	入口分流器
outlet collector	出口集水器
quiet zone	滞留区
perpendicular to	垂直于
sand drain	排砂
back – pressure	背压
atmospheric vent system	常压放空系统
short circuiting	短路

Language Focus

1. In producing operations it is often necessary to handle wastewater that may include water produced with crude oil, rain water, and washdown water.

（参考译文：生产操作中，常常需要处理各种各样的废水，其中包括随原油采出的污水、雨水和冲洗水。）

本句中"it"是形式主语，真正的主语是"to handle wastewater that may include water produced with crude oil, rain water, and washdown water"；"that"引导限制性定语从句，指代"wastewater"；"produced with crude oil"是过去分词短语作后置定语修饰"water"。

2. In most onshore locations the water cannot be disposed of on the surface, due to possible salt contamination, and must be injected into an acceptable disposal formation or disposed of by evaporation.

（参考译文：大多数陆上油区的污水不能直接在地面排放，因为污水可能使土壤受到盐侵污染，必须通过污水井注入适当地层或者通过地面蒸发处理。）本句中介词短语"due to"意为"因为、由于"；"and"连接两个并列的谓语"cannot be disposed of..."和"must be injected into..."，"must"后又有两个"or"连接的并列动词"be injected into"和"disposed of"，后者前省略了"be"。

3. For safety considerations, closed drains, if they exist in the process, should never be tied into atmospheric drains and should be routed to a pressure vessel prior to entering an atmospheric tank or pile.

（参考译文：出于安全考虑，如果设备中有排水暗管，应先接入压力容器，再进常压罐或排放管桩。）

本句中"For safety considerations"是目的状语；"closed drains should never be..."是主谓结构，中间插入了"if"引导的条件状语从句"they exist in the process"，"they"指代"closed drains"；"prior to"意为"在……前面"。

4. Internal waves due to surging in horizontal vessels can trigger a level float even though the volume of liquid between the normal operating level and the high level shutdown is equal to or larger than that in a vertical vessel.

（参考译文：而在卧式容器中，尽管正常操作液面与高液面停机液面之间的液体体积与立式容器相等或更大些，但脉动引起的内部波浪会使液面浮子误触动停机机构。）

本句中的"due to"介词短语表示"原因"，"even though"引导让步状语从句"the volume of liquid between the normal operating level and the high level shutdown is equal to or larger than that in a vertical vessel"，"that"指代"the volume of liquid between the normal operating level and the high level"。

🔲 Reinforced Learning

Ⅰ. Answer the following questions for a comprehension of the text.

1. What wastewater should we handle in producing operations?

2. What is the maximum hydrocarbon content in the water?

3. What is the purpose of this article?

4. Where is produced water disposed of offshore and onshore?

5. When is the pressure vessel used instead of atmospheric skimmers?

Ⅱ. Multiple choice：choose the correct one from the alternative answers to give the exact meaning of the words.

1. In most onshore locations the water cannot be disposed of on the surface, due to possible salt contamination, and must be injected into an acceptable disposal formation or disposed of by evaporation.

A. containment
B. inclusion

C. pollution
D. configuration

2. When this design procedure is followed, the engineer will be able to develop a process flowsheet, determine equipment sizes, and evaluate vendor pro-

posals for any wastewater treating system.

    A. supplier   B. designer   C. engineer   D. manager

    3. Depending upon the severity of the treating problem, secondary treating utilizing a CPI, crossflow separator, or a flotation unit may be required.

    A. seriousness   B. difficulty   C. stiffness   D. possibility

    4. Offshore, produced water can be piped directly overboard after treating, or it can be routed through a disposal pile or a skim pile.

    A. disconnection   B. entrance   C. separation   D. release

    5. In the quiet zone between the spreader and the water collector, some coalescence can occur and the buoyancy of the oil droplets causes them to rise counter to the water flow.

    A. tranquil   B. quite   C. peace   D. remain

    6. In horizontal skimmers the oil droplets rise perpendicular to the flow of the water, as shown in Figure 2.3.2.

    A. vertical   B. straight   C. horizontal   D. angular

    7. Vertical vessels are less susceptible to high level shutdowns due to liquid surges.

    A. doubtable   B. possible   C. easy   D. separable

    8. Potential gas blowby through the upstream vessel dump system could create too much back-pressure in an atmospheric vent system.

    A. input   B. inject   C. steam   D. leakage

    9. A minimum residence time of 10 to 30 minutes should be provided to assure that surges do not upset the system and to provide for some coalescence.

    A. minus   B. least   C. most   D. more

    10. Skimmers with long residence times require baffles to attempt to distribute the flow and eliminate short circuiting.

    A. circus   B. suburbia   C. electricity line   D. boundary

    III. Multiple choice: read the four suggested translations and choose the best answer.

    1. In most onshore locations the water cannot be disposed of on the surface, due to possible salt contamination, and must be injected into an acceptable disposal formation or disposed of by evaporation.

    A. 形成   B. 发展   C. 层级   D. 地层

2. When this design procedure is followed, the engineer will be able to develop a process flowsheet, determine equipment sizes, and evaluate vendor proposals for any wastewater treating system.

　　A. 流水线　　　　B. 草图　　　　　C. 工艺流程图　　D. 工艺步骤

3. When this design procedure is followed, the engineer will be able to develop a process flowsheet, determine equipment sizes, and evaluate vendor proposals for any wastewater treating system.

　　A. 商贩　　　　　B. 承包商　　　　C. 施工方　　　　D. 供应商

4. Offshore, produced water can be piped directly overboard after treating, or it can be routed through a disposal pile or a skim pile.

　　A. 排放管道　　　B. 隔油管桩　　　C. 脱脂管桩　　　D. 排污管道

5. Deck drains must be treated for removal of "free" oil. Onshore, the water will normally be reinjected in the formation or be pumped into a disposal well.

　　A. 处理井　　　　B. 清理井　　　　C. 污水井　　　　D. 分离井

6. For safety considerations, closed drains, if they exist in the process, should never be tied into atmospheric drains and should be routed to a pressure vessel prior to entering an atmospheric tank or pile.

　　A. 常压罐　　　　B. 大气槽　　　　C. 气压槽　　　　D. 储气罐

7. Some vertical skimmers have inlet spreaders and outlet collectors to help even the distribution of the flow, as shown in Figure 2.3.1.

　　A. 进口分离器 排出收集器　　　　B. 进入分散器 排出收水器
　　C. 入口分流器 出口集水器　　　　D. 入口散发器 出口收集器

8. In the quiet zone between the spreader and the water collector, some coalescence can occur and the buoyancy of the oil droplets causes them to rise counter to the water flow.

　　A. 安静　　　　　B. 平流　　　　　C. 遗留　　　　　D. 滞留

9. Potential gas blowby through the upstream vessel dump system could create too much back-pressure in an atmospheric vent system.

　　A. 常压防空系统　　　　　　　　　B. 通风系统
　　C. 空控体制　　　　　　　　　　　D. 气压控制体制

10. Skimmers with long residence times require baffles to attempt to distribute the flow and eliminate short circuiting.

　　A. 短路　　　　　B. 断电　　　　　C. 断路　　　　　D. 短电

IV. Put the following sentences into Chinese.

1. In offshore areas where discharge to the sea is allowed, the governing regulatory body specifies the maximum hydrocarbon content in the water that may be discharged overboard.

2. The choice of pressure versus atmospheric vessel for the skimmer tank is not determined solely by the water treating requirements. The overall needs of the system need to be considered in this decision.

3. For safety considerations, closed drains, if they exist in the process, should never be tied into atmospheric drains and should be routed to a pressure vessel prior to entering an atmospheric tank or pile.

4. Horizontal vessels are more efficient at water treating because the oil droplets do not have to flow countercurrent to the water flow.

5. Internal waves due to surging in horizontal vessels can trigger a level float even though the volume of liquid between the normal operating level and the high level shutdown is equal to or larger than that in a vertical vessel.

V. Put the following paragraphs into Chinese.

1. The simplest form of primary treating equipment is a settling tank or vessel. These items are normally designed to provide long residence times during which coalescence and gravity separation can occur. Skimmers can be either vertical or horizontal in configuration. In vertical skimmers the oil droplets must rise upward countercurrent to the downward flow of the water. Small amounts of gas liberated from the water help to "float" the oil droplets. In the quiet zone between the spreader and the water collector, some coalescence can occur and the buoyancy of the oil droplets causes them to rise counter to the water flow. Oil will be collected and skimmed off the surface.

2. The potential benefits of providing much more residence time will probably not be cost efficient beyond this point. Skimmers with long residence times require baffles to attempt to distribute the flow and eliminate short circuiting. Tracer studies have shown that skimmer tanks, even those with carefully designed spreaders and baffles, exhibit poor flow behavior and short circuiting. This is probably due to density and temperature differences, deposition of solids, corrosion of spreaders, etc.

2.4　Produced – Water Treating Systems (2)

Guidance to Reading

The quality of produced water varies significantly based on the geochemistry of the producing formation, the type of hydrocarbon produced, and the characteristics of the producing well. Treatment is required to improve the quality of produced water so that it can be put to beneficial use. If produced water meets appropriate water quality criteria, it may be used beneficially for purposes such as irrigation, livestock watering, and municipal and industrial uses. The text presented the design procedure to be followed by engineers to develop a process flowsheet, determine equipment sizes, and evaluate vendor proposals for any wastewater treating system.

Text

Plate Coalescers

Plate coalescers are skim tanks or vessels that use internal plates to improve the gravity separation process. Various configurations of plate coalescers have been devised. These are commonly called **parallel plate interceptors (PPI)**, **corrugated plate interceptors (CPI)**, or **cross – flow separator**s. All of these depend on gravity separation to allow the oil droplets to rise to a plate surface where coalescence and capture occur. As shown in Figure 2.4.1 flow is split between a number of parallel plates spaced a short distance apart. To facilitate capture of the oil droplet the plates **are inclined to** the horizontal.

Disposal Piles

Disposal piles are large diameter (24 – to 48 – inch) **open – ended pipe**s attached to the platform and extending below the surface of the water. Their main uses are to (1) concentrate all **platform discharge**s into one location, (2) provide a **conduit** protected from wave action so that discharges

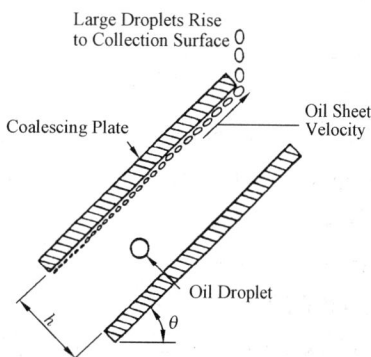

Figure 2.4.1　Plate coalescers

can be placed deep enough to prevent **sheens from occurring during upset conditions, and** (3) provide an alarm or shutdown point **in the event of** a failure causing oil to flow overboard.

Most authorities having jurisdiction require all produced water to be treated (**skimmer tank, coalescer**, or **flotation**) prior to disposal in a disposal pile. In some locations, disposal piles are permitted to collect treated produced water, treated sand, liquids from **drip pan**s and **deck drain**, and as a final trap for hydrocarbon liquids in the event of equipment upsets.

Disposal piles are particularly useful for deck **drainage** disposal. This flow, which originates either from rainwater or washdown water, typically contains oil droplets dispersed in an oxygen – laden fresh or saltwater phase. The oxygen in the water makes it highly corrosive and **commingling** with produced water may lead to **scale deposition** and **plugging** in skimmer tanks, plate coalescers, or flotation units. The flow is highly irregular and would thus cause upsets throughout these devices. Finally, this flow must **gravitate** to a low point for collection and either be pumped up to a higher level for treatment or treated at that low point. Disposal piles are excellent for this purpose. They can be protected from corrosion, they are by design located low enough on the platform to **eliminate** the need for pumping the water, they are not severely affected by large **instantaneous** flow rate changes (**effluent** quality may be affected to some extent but the operation of the pile can continue), they contain no small passages **subject to** plugging by scale buildup, and they minimize commingling with the process since they are the last piece of treating equipment before disposal.

The disposal pile length should be as long as the water depth permits in shallow water to provide for maximum oil **containment** in the event of a **malfunction** and to minimize the potential appearance of any sheen. In deep water, the length is set to assure that an alarm and then a shutdown signal can be measured before the pile fills with oil. These signals must be high enough so as not to register tide changes.

Skim Pile

The skim pile is a type of disposal pile. As shown in Figure 2.4.2, flow through the multiple series of baffle plates creates zones of no flow that reduce the distance a given oil droplet must rise to be separated from the main flow.

Once in this zone, there is plenty of time for coalescence and gravity separation. The larger droplets then migrate up the underside of the baffle to an oil collection system.

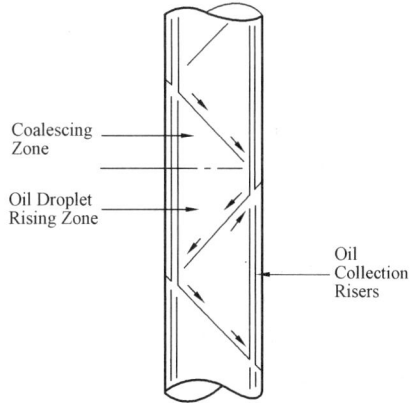

Figure 2.4.2 Skim pile flow pattern

Besides being more efficient than standard disposal piles, from an oil separation standpoint, skim piles have the added benefit of providing for some degree of sand cleaning. Most authorities having jurisdiction state that produced sand must be disposed of without "free oil." It is doubtful that sand from a vessel drain meets this criterion when disposed of in a standard disposal pile.

Sand **traversing** the length of a skim pile will **abrade** on the baffles and be water washed. This can be said to remove the free oil that is then captured in a **quiescent zone**.

Equipment selection procedure

Federal regulations and most authorities having jurisdiction require that produced water from the free – water knockout receive at least some form of primary treatment before being sent to a disposal pile or skim pile. Deck drainage may be routed to a properly sized disposal pile that will remove "free oil."

Every water treating system design must be with the sizing, for liquid separation of a free – water knockout, heater treater, or three – phase separator.

With the exception of these restraints the design engineer is free to arrange the system as he **sees fit**. There are many potential combinations of the equipment previously described. Under a certain set of circumstances, it may be appropriate to dump the water from a free – water knockout directly to a skim tank for final treatment before discharge. Under other circumstances, a full system of plate coalescers, flotation units, and skim piles may be needed. In the final analysis the choice of a particular combination of equipment and their sizing must rely rather heavily on the judgment and experience of the design engineer. The following procedure is meant only as a guideline and not as a substitute for his judgment and experience. In no instance is this procedure meant to be used without proper weight given to operational experience in the specific area.

(1)Determine the oil content of the produced water **influent**.

(2)Determine dispersed oil effluent quality.

(3) Determine oil drop size distribution in the influent produced water stream.

(4)Determine oil particle diameter that must be treated to meet effluent quality required.

(5)Determine skimmer dimensions.

(6)Determine plate coalescer dimensions.

(7)Choose skim tank, **SP Pack**, or plate coalescer for application, considering cost and space available.

(8)Choose method of handling deck drainage.

Words and Expressions

conduit	导管
sheen	水面油污
coalescer	聚结器
flotation	浮选设备
drainage	污水
commingle	混合
plug	堵塞
gravitate	在重力作用下
eliminate	消除
instantaneous	瞬间的
effluent	污水
containment	容量
malfunction	故障
traverse	穿过
abrade	摩擦
influent	流入的

Phrases and Expressions

plate coalescer	板式聚结器
parallel plate interceptor（PPI）	平行板隔油器

corrugated plate interceptor（CPI）	波纹板隔油器
cross – flow separator	交叉流分流器
be inclined to	倾斜
disposal pile	污水排放管桩
open – ended pipe	敞口管
platform discharge	平台排放的污水
in the event of	万一
skimmer tank	隔油罐
drip pan	油滴收集盘
deck drain	甲板冲洗水
scale deposition	结垢
subject to	受制于
skim pile	隔油管桩
quiescent zone	静止区
see fit	认为……是恰当的
SP Pack	盘管组件

Language Focus

1. All of these depend on gravity separation to allow the oil droplets to rise to a plate surface where coalescence and capture occur.

（参考译文:所有这些都是根据重力分离原理使油滴浮升到板表面,在此进行聚结和油滴捕集。）

本句中"where"引导限制性定语从句"coalescence and capture occur",指代"in the plate surface"。

2. Finally,this flow must gravitate to a low point for collection and either be pumped up to a higher level for treatment or treated at that low point.

（参考译文:最终,这些污水在重力作用下会流到一个较低部位集中起来,如果处理设备标高较高,应当用泵;如果处理设备标高较低,可以不用泵。）

本句中"and"连接两个并列的动词短语"gravitate to a low point for collection"和"either be pumped up to a higher level for treatment or treated at that low point";连词搭配"either… or…"表示"两者中任意一个","either"和"or"后面的结构一致,"or"后省略了"be"。

3. Besides being more efficient than standard disposal piles, from an oil

separation standpoint, skim piles have the added benefit of providing for some degree of sand cleaning.

（参考译文：隔油管桩比标准排放管桩能更有效地分离油，而且，从油分离的角度看，隔油管桩还有净化油污砂子的优点。）

本句中"besides"作介词，意为"除……之外"；"from...standpoint"意为"从……角度来看"。

⬚ Reinforced Learning

Ⅰ. Answer the following questions for a comprehension of the text.

1. What are various configurations of plate coalescers?

2. What are the main uses of disposal piles?

3. What role can the disposal pile play in the event of a malfunction?

4. What benefits do skim piles have?

5. What are procedures for equipment selection?

Ⅱ. Multiple choice: choose the correct one from the alternative answers to give the exact meaning of the words.

1. To facilitate capture of the oil droplet the plates are inclined to the horizontal.

　　A. prone　　　　　B. level　　　　　C. straight　　　　　D. horizontal

2. Their main uses are to concentrate all platform discharges into one location.

　　A. expulsion　　　B. exchange　　　C. incorporation　　　D. consumption

3. Provide a conduit protected from wave action so that discharges can be placed deep enough to prevent sheens from occurring during upset conditions.

　　A. conduct　　　　B. outlet　　　　C. channel　　　　　D. barrel

4. Provide an alarm or shutdown point in the event of a failure causing oil to flow overboard.

　　A. on occasion of　　　　　　　B. in case of

　　C. in terms of　　　　　　　　D. in spite of

5. Most authorities having jurisdiction require all produced water to be treated (skimmer tank, coalescer, or flotation) prior to disposal in a disposal pile.

　　A. jury　　　　　B. court　　　　　C. legislation　　　　D. rights

6. The oxygen in the water makes it highly corrosive and commingling with

produced water may lead to scale deposition and plugging in skimmer tanks, plate coalescers, or flotation units.

    A. producing   B. leaving   C. mixing   D. remaining

    7. They are not severely affected by large <u>instantaneous</u> flow rate changes.

    A. momentary   B. consecutive   C. successive   D. alternative

    8. The disposal pile length should be as long as the water depth permits in shallow water to provide for maximum oil containment in the event of a <u>malfunction</u> and to minimize the potential appearance of any sheen.

    A. operation   B. trouble   C. manufacture   D. destruction

    9. Sand traversing the length of a skim pile will <u>abrade</u> on the baffles and be water washed.

    A. flow   B. cross   C. scrape   D. run

    10. Deck <u>drainage</u> may be routed to a properly sized disposal pile that will remove "free oil".

    A. water   B. steam   C. oil   D. sewage

Ⅲ. Multiple choice: read the four suggested translations and choose the best answer.

    1. Various configurations of plate coalescers have been devised. These are commonly called parallel plate interceptors (PPI), <u>corrugated plate interceptors (CPI)</u>, or cross – flow separators.

    A. 平行板隔油器   B. 交叉流分流器
    C. 波纹板隔油器   D. 板式聚结器

    2. Disposal piles are large diameter <u>open – ended pipes</u> attached to the platform and extending below the surface of the water.

    A. 敞口管   B. 开放管   C. 无尾管   D. 分口管

    3. Most <u>authorities</u> having jurisdiction require all produced water to be treated prior to disposal in a disposal pile.

    A. 权威   B. 部门   C. 专家   D. 环保机构

    4. This flow, which originates either from rainwater or washdown water, typically contains oil droplets dispersed in an <u>oxygen – laden fresh or saltwater</u> phase.

    A. 不含氧气的淡水或盐水   B. 少量氧气的淡水或盐水
    C. 富含氧气的淡水或盐水   D. 含有氧气的淡水或盐水

    5. They contain no small passages <u>subject to</u> plugging by scale buildup.

A. 题目是　　　B. 受制于　　　C. 主题是　　　D. 科目是

6. The disposal pile length should be as long as the water depth permits in shallow water to provide for maximum oil <u>containment</u> in the event of a malfunction and to minimize the potential appearance of any sheen.

A. 包含　　　B. 容量　　　C. 含量　　　D. 容度

7. This can be said to remove the free oil that is then captured in a <u>quiescent zone</u>.

A. 静止区　　　B. 安静区　　　C. 静音区　　　D. 专属区

8. With the exception of these restraints the design engineer is free to arrange the system as he <u>sees fit</u>.

A. 正好看见　　　　　　　B. 看见……是适合的

C. 考虑……可以实施　　　　D. 认为……是恰当的

9. Determine the oil content of the produced water <u>influent</u>.

A. 影响的　　　B. 流入的　　　C. 充足的　　　D. 宽广的

10. Choose skim tank, <u>SP Pack</u>, or plate coalescer for application, considering cost and space available.

A. 隔油管桩　　　B. 浮选设备　　　C. 盘管组件　　　D. 排放管桩

IV. Put the following sentences into Chinese.

1. All of these depend on gravity separation to allow the oil droplets to rise to a plate surface where coalescence and capture occur.

2. Their main uses are to (1) concentrate all platform discharges into one location, (2) provide a conduit protected from wave action so that discharges can be placed deep enough to prevent sheens from occurring during upset conditions, and (3) provide an alarm or shutdown point in the event of a failure causing oil to flow overboard.

3. The disposal pile length should be as long as the water depth permits in shallow water to provide for maximum oil containment in the event of a malfunction and to minimize the potential appearance of any sheen.

4. Flow through the multiple series of baffle plates creates zones of no flow that reduce the distance a given oil droplet must rise to be separated from the main flow.

5. Federal regulations and most authorities having jurisdiction require that produced water from the free – water knockout receive at least some form of primary treatment before being sent to a disposal pile or skim pile.

V. Put the following paragraphs into Chinese.

1. Besides being more efficient than standard disposal piles, from an oil separation standpoint, skim piles have the added benefit of providing for some degree of sand cleaning. Most authorities having jurisdiction state that produced sand must be disposed of without "free oil." It is doubtful that sand from a vessel drain meets this criterion when disposed of in a standard disposal pile.

2. The deck drainage, which originates either from rainwater or washdown water, typically contains oil droplets dispersed in an oxygen – laden fresh or saltwater phase. The oxygen in the water makes it highly corrosive and commingling with produced water may lead to scale deposition and plugging in skimmer tanks, plate coalescers, or flotation units. The flow is highly irregular and would thus cause upsets throughout these devices. Finally, this flow must gravitate to a low point for collection and either be pumped up to a higher level for treatment or treated at that low point.

Chapter 3   Natural Gas Origin and Transportation

3.1   Natural Gas Origin and Composition (1)

🔲 Guidance to Reading

*Natural gas has been formed by the degradation of organic matter accumulated in the past millions of years, existing under pressure in rock reservoirs in the Earth's crust, either in conjunction with and dissolved in heavier hydrocarbons and water or by itself. Natural gas is a hydrocarbon gas mixture consisting primarily of **methane**, but commonly includes varying amounts of other higher alkanes and even a lesser percentage of **carbon dioxide**, **nitrogen**, and **hydrogen sulfide**. Natural gas can be broadly categorized **into nonassociated gas**, associated gas and **unconventional** gas, as well as into **lean** or rich, wet or dry.*

🔲 Text

Two main mechanisms (**biogenic** and **thermogenic**) are responsible for this degradation. **Biogenic gas** is formed at shallow depths and low temperatures by the **anaerobic** bacterial decomposition of **sedimentary organic matter**. In contrast, thermogenic gas is formed at deeper depths by (1) thermal cracking of sedimentary organic matter into hydrocarbon liquids and gas (this gas is cogenetic with oil and is called "primary" thermogenic gas) and (2) thermal cracking of oil at high temperatures into gas ("secondary" thermogenic gas) and **pyrobitumen**. Biogenic gas consists almost entirely of methane. In contrast, thermogenic gas can also contain significant concentrations of **ethane, propane, butanes**, and heavier hydrocarbons.

The principal constituent of natural gas is methane. Other constituents are **paraffinic hydrocarbons** such as ethane, propane, and the butanes. Many natural gases contain nitrogen as well as carbon dioxide and hydrogen sulfide. Trace quantities of **argon**, hydrogen, and **helium** may also be present. The composition of natural gas can vary widely. Natural gas can also contain a small proportion of C + 5 hydrocarbons. When separated, this fraction is a light gasoline. Some **aromatics** such as **benzene, toluene**, and **xylenes** can also be present, raising safety issues due to their **toxicity**. **Mercury** can also be present either as a metal in vapor phase or as an **organometallic compound** in liquid fractions. Concentration levels are generally very small, but even at very small concentration levels, mercury can be **detrimental** due its toxicity and its **corrosive properties**.

Natural gas is considered "dry" when it is almost pure methane, having had most of the other commonly associated hydrocarbons removed. When other hydrocarbons are present, the natural gas is "wet". The composition of natural gas varies depending on the field, formation, or reservoir from which it is **extracted**. Since the composition of natural gas is never constant, there are standard test methods by which the composition of natural gas can be determined and thus prepared for use.

Gas Sources

Natural gas produced from geological formations comes in a wide array of compositions. The varieties of gas compositions can be broadly categorized into three distinct groups: (1) nonassociated gas that occurs in **conventional** gas fields, (2) associated gas that occurs in conventional oil fields, and (3) **unconventional** gas. Some types of unconventional gas resources include "tight gas" or "tight sands gas," which is found in **low – permeability rock**; "**coalbed methane**," which is natural gas that has been formed along with the geological processes that formed coal; "**gas hydrates**," which are ice – like structures of water and gas located under the **permafrost**; and "deep gas," which is found at levels much deeper than conventional gas. Of the unconventional gas sources, the one most important to the gas transportation industry is coalbed methane.

Associated Gas

Associated gas is produced during crude oil production and is the gas that is associated with crude oil. Crude oil cannot be produced without producing some of its associated gas, which comes out of solution as the pressure is reduced on the way to and on the surface. Crude oil in the reservoir with minimal or no dissolved associated gas is rare.

After the production fluids are brought to the surface, they are separated at a tank battery at or near the production lease into a hydrocarbon liquid stream (crude oil or *gas condensate*), a produced water stream (**brine** or salty water), and a **gaseous stream**. The gaseous stream is traditionally very rich (*rich gas*) *in natural gas liquids* (NGLs). Natural gas liquids include ethane, propane, butanes, and pentanes and higher molecular weight hydrocarbons (C_{6+}). The higher molecular weight hydrocarbons product is commonly referred to as *natural gasoline*.

When referring to natural gas liquids in the natural gas stream, the term "gal/ft^3" is used as a measure of hydrocarbon richness. Depending on its content of heavy components, natural gas can be considered as rich (5 or 6 gallons or more of recoverable hydrocarbons per cubic feet) or lean (less than 1 gallon of recoverable hydrocarbons per cubic feet). However, the terms *rich gas* and *lean gas*, as used in the gas – processing industry, are not precise indicators of gas quality but only indicate the relative amount of natural gas liquids in the gas stream.

In the case of associated gas, crude oil may be assisted up the well bore by gas lift. Thus, gas is compressed into the **annulus** of the well and then injected by means of a gas lift valve near the bottom of the well into the crude oil column in the tubing. At the top of the well the crude oil and gas mixture passes into a separation plant that drops the pressure down to nearly atmospheric in two stages. The crude oil and water exit the bottom of the lower pressure separator, from where it is pumped to tanks for separation of the crude oil and water. The gas produced in the separators and the gas that comes out of solution with the produced crude oil (surplus gas) are then treated to separate out the NGLs that are treated in a gas plant to provide propane and butane or a mixture of the two (**liquefied petroleum gas**). The higher boiling residue, after the propane and butane are removed, is **condensate**, which is mixed with the crude oil or exported as a separate product. The gas itself is then *dry* and, after compression, is suitable to be injected into the natural gas system where it substitutes for natural gas from the nonassociated gas reservoir. Pretreated associated gas from other fields enters the system at this stage. Another use for the gas is as fuel for the gas turbines on site. This gas is treated in a fuel gas plant to ensure that it is clean and at the correct pressure. The start – up fuel gas supply will be from the main gas system, but facilities exist to collect and treat low – pressure gas from the various other plants as a more economical fuel source.

Worldwide, governments are mandating that producers stop flaring associated gas, as their citizens perceive that it is a waste of a valuable nonrenewable resource. There are often regulatory restrictions on when produced gas can be reinjected, or flared, with an understanding that any reinjected gas must eventually be produced.

Words and Expressions

methane	甲烷
nitrogen	氮
unconventional	非常规的
lean	贫乏的
biogenic	源于生物的
thermogenic	热源的
anaerobic	厌氧的
pyrobitumen	焦沥青
ethane	乙烷
propane	丙烷
butane	丁烷
argon	氩
helium	氦
aromatic	芳香族化合物
benzene	苯
toluene	甲苯
xylene	二甲苯
toxicity	毒性
mercury	汞
detrimental	有害的
extract	提取;开采
permafrost	永久冻土
brine	盐水
annulus	环
condensate	凝析液

🔲 Phrases and Expressions

carbon dioxide	二氧化碳
hydrogen sulfide	硫化氢
nonassociated gas	非伴生气
biogenic gas	生物气
sedimentary organic matter	沉积有机质
paraffinic hydrocarbon	烷烃
organometallic compound	有机金属化合物
corrosive property	腐蚀性
low – permeability rock	低渗透致密砂岩
coalbed methane	煤层气
gas hydrate	天然气水合物
gaseous stream	气流
liquefied petroleum gas	液化石油气

🔲 Language Focus

1. Natural gas is considered "dry" when it is almost pure methane, having had most of the other commonly associated hydrocarbons removed.

（参考译文：当天然气脱除大部分其他常见的伴生烃,组分几乎接近于纯甲烷时,就称为"干气"。）

本句中"when"引导时间状语从句"it is almost pure methane",后面的现在分词短语"having had most of the other commonly associated hydrocarbons removed"作状语,表示先一步的动作。

2. Crude oil cannot be produced without producing some of its associated gas, which comes out of solution as the pressure is reduced on the way to and on the surface.

（参考译文：原油开采过程中不可能只采油不采气,气体总是会随着井筒压力的下降而从原油中析出。）

本句中"which"引导非限制性定语从句,指代前面的 "associated gas";"as"引导时间状语从句"the pressure is reduced on the way to and on the surface"。

3. The gas itself is then *dry* and, after compression, is suitable to be injected into the natural gas system where it substitutes for natural gas from the nonasso-

ciated gas reservoir.

（参考译文：此时的气体就是干气了，经过压缩，就可以输入天然气系统，顶替非伴生气。）

本句中"and"连接两个并列的部分"*dry*"和"suitable"；"where"引导限制性定语从句"it substitutes for natural gas from the nonassociated gas reservoir"，指代"in the natural gas system"；"substitutes for"的意思是"替代"。

Reinforced Learning

Ⅰ. Answer the following questions for a comprehension of the text.

1. What are the mechanisms of the degradation of organic matter?

2. What does natural gas consist of?

3. What are sources of natural gas?

4. How is nonassociated gas produced?

5. Why should associated gas flaring be stopped?

Ⅱ. Multiple choice: choose the correct one from the alternative answers to give the exact meaning of the words.

1. Natural gas exists in nature under pressure in rock reservoirs in the Earth's crust, either in conjunction with and dissolved in heavier hydrocarbons and water or by itself.

A. cluster　　　B. basin　　　C. group　　　D. sandstone

2. Natural gas has been formed by the degradation of organic matter accumulated in the past millions of years.

A. development　B. formation　C. construction　D. reduction

3. The principal constituent of natural gas is methane.

A. character　　B. matter　　C. components　D. gasoline

4. Some aromatics such as benzene, toluene, and xylenes can also be present, raising safety issues due to their toxicity.

A. venom　　　B. element　　C. value　　　D. weight

5. Concentration levels are generally very small, but even at very small concentration levels, mercury can be detrimental due its toxicity and its corrosive properties.

A. harmful　　B. destructive　C. effective　　D. functional

6. The composition of natural gas varies depending on the field, formation,

or reservoir from which it is <u>extracted</u>.

    A. produced   B. exploited   C. deposited   D. dragged

    7. After the production fluids are brought to the surface, they are separated at a tank battery at or near the production lease into a hydrocarbon liquid stream (crude oil or gas condensate), a produced water stream (<u>brine</u>), and a gaseous stream.

    A. salty water   B. preserved water

    C. mineral water   D. gas liquid

    8. Depending on its content of heavy components, natural gas can be considered as rich (5 or 6 gallons or more of recoverable hydrocarbons per cubic feet) or <u>lean</u> (less than 1 gallon of recoverable hydrocarbons per cubic feet).

    A. naught   B. abundant   C. none   D. skinny

    9. The higher boiling residue, after the propane and butane are removed, is <u>condensate</u>, which is mixed with the crude oil or exported as a separate product.

    A. NGL   B. LPG   C. COS   D. CBM

    10. Worldwide, governments are mandating that producers stop flaring associated gas, as their citizens perceive that it is a waste of a valuable <u>nonrenewable</u> resource.

    A. new   B. unregenerate   C. unavailable   D. productive

    III. Multiple choice: read the four suggested translations and choose the best answer.

    1. Natural gas exists in nature under pressure in rock reservoirs in the Earth's crust, either in <u>conjunction with</u> and dissolved in heavier hydrocarbons and water or by itself.

    A. 联系   B. 关联   C. 连同   D. 同时

    2. Natural gas has been formed by the <u>degradation</u> of organic matter accumulated in the past millions of years.

    A. 破坏   B. 降解   C. 分离   D. 掩埋

    3. Biogenic gas is formed at shallow depths and low temperatures by the anaerobic bacterial decomposition of <u>sedimentary</u> organic matter.

    A. 沉积的   B. 火成的   C. 变质的   D. 石灰的

    4. In contrast, thermogenic gas is formed at deeper depths by (1) thermal cracking of sedimentary organic matter into hydrocarbon liquids and gas (this gas is cogenetic with oil and is called "primary" thermogenic gas) and

（2）thermal cracking of oil at high temperatures into gas（"secondary" thermogenic gas）and pyrobitumen.

　　A. 重烃　　　　B. 有机质　　　C. 焦沥青　　　D. 厌氧菌

5. In contrast, thermogenic gas can also contain significant concentrations of ethane, propane, butanes, and heavier hydrocarbons.

　　A. 甲烷、乙烷、丙烷以及重氢　　B. 氩、氢、氦以及碳氢
　　C. 乙烷、丙烷、丁烷以及重烃　　D. 苯、甲苯、二甲苯以及重氢

6. Gas geochemistry readily reveals whether a gas is biogenic or thermogenic.

　　A. 二基因或多基因　　　　　　B. 生物基因或非生物基因
　　C. 生物制造还是天然制造　　　D. 生物成因还是热成因

7. Many natural gases contain nitrogen as well as carbon dioxide and hydrogen sulfide.

　　A. 氮气　　　　B. 硫化氢　　　C. 碳化氢　　　D. 二氧化碳

8. Some aromatics such as benzene, toluene, and xylenes can also be present, raising safety issues due to their toxicity.

　　A. 氩、氢、氦　　　　　　　　B. 甲烷、乙烷、丙烷
　　C. 苯、甲苯、二甲苯　　　　　D. 乙烷、丙烷、丁烷

9. The varieties of gas compositions can be broadly categorized into three distinct groups：（1）nonassociated gas that occurs in conventional gas fields,（2）associated gas that occurs in conventional oil fields, and（3）unconventional gas.

　　A. 非传统的　　B. 不正常的　　C. 无惯例的　　D. 非常规的

10. Associated gas is produced during crude oil production and is the gas that is associated with crude oil.

　　A. 联合气　　　B. 天然气　　　C. 联产气　　　D. 伴生气

IV. Put the following sentences into Chinese.

1. Natural gas exists in nature under pressure in rock reservoirs in the Earth's crust, either in conjunction with and dissolved in heavier hydrocarbons and water or by itself.

2. In contrast, thermogenic gas is formed at deeper depths by（1）thermal cracking of sedimentary organic matter into hydrocarbon liquids and gas（this gas is cogenetic with oil and is called "primary" thermogenic gas）and（2）thermal cracking of oil at high temperatures into gas（"secondary" thermo-

genic gas) and pyrobitumen.

3. Some types of unconventional gas resources include "tight gas" or "tight sands gas", which is found in low – permeability rock; "coalbed methane", which is natural gas that has been formed along with the geological processes that formed coal; "gas hydrates", which are ice – like structures of water and gas located under the permafrost; and "deep gas", which is found at levels much deeper than conventional gas.

4. However, the terms *rich gas* and *lean gas*, as used in the gas – processing industry, are not precise indicators of gas quality but only indicate the relative amount of natural gas liquids in the gas stream.

5. There are often regulatory restrictions on when produced gas can be reinjected, or flared, with an understanding that any reinjected gas must eventually be produced.

V. Put the following paragraphs into Chinese.

1. Natural gas is considered "dry" when it is almost pure methane, having had most of the other commonly associated hydrocarbons removed. When other hydrocarbons are present, the natural gas is "wet".

2. In the case of associated gas, crude oil may be assisted up the well bore by gas lift. Thus, gas is compressed into the annulus of the well and then injected by means of a gas lift valve near the bottom of the well into the crude oil column in the tubing. Another use for the gas is as fuel for the gas turbines on site. This gas is treated in a fuel gas plant to ensure that it is clean and at the correct pressure. The start – up fuel gas supply will be from the main gas system, but facilities exist to collect and treat low – pressure gas from the various other plants as a more economical fuel source.

3.2   Natural Gas Origin and Composition (2)

［ Guidance to Reading

*Associated gas is produced during crude oil production in the conventional oil and gas reservoir while non – associated gas is produced in the conventional gas reservoir. Coalbed methane is extracted from coal beds, referring to a form of natural gas absorbed into the solid **matrix** of the coal. Shale gas and **tight gas** are characterized by low permeability, with more limited ability of gas to* *flow through the rock than is the case with a conventional reservoir. Unlike conventional hydrocarbon reservoirs, these gases often remain trapped in, or close to, its source rock. The natural gas cannot be sold before appropriate treatment for acid gas reduction and **odorization** to ensure safety.*

🔲 Text

Nonassociated Gas

Nonassociated gas (sometimes called "gas well gas") is produced from geological formations that typically do not contain much, if any, higher boiling hydrocarbons (gas liquids) than methane. Nonassociated gas can contain nonhydrocarbon gases such as carbon dioxide and hydrogen sulfide. Nonassociated gas is directly controllable by the producer; one just turns the valves. The gas flows up the well under its own energy, through the wellhead control valves, and along the flow line to the treatment plant. Treatment requires the temperature of the gas to be reduced to a point dependent upon the pressure in the pipeline so that all liquids that would exist at pipeline temperature and pressure condense and are removed.

Unconventional Gas

Unconventional gas refers to a part of the gas resource base that has traditionally been considered difficult or costly to produce. In this report, we focus on the three main categories of unconventional gas:

Shale Gas

Shale gas, one of the most rapidly growing forms of natural gas, is defined as natural gas from shale formations. The shale acts as both the source and the reservoir for the natural gas. Shale formations are characterised by low permeability, with more limited ability of gas to flow through the rock than is the case with a conventional reservoir. Older shale gas wells were vertical while more recent wells are primarily horizontal and need **artificial stimulation**, like **hydraulic fracturing**, to produce. Only shale formations with certain characteristics will produce gas. In large measure this is attributable to significant advances in the use of **horizontal drilling** and well stimulation technologies and refinement in the cost – effectiveness of these technologies. Hydraulic fracturing is the most significant of these.

Tight Gas

Tight gas is a general term for natural gas found in low permeability formations. Generally, we classify as tight gas those low permeability gas reservoirs that cannot produce economically without the use of technologies to stimulate flow of the gas towards the well, such as hydraulic fracturing. Tight gas is often a poorly defined category with no clear boundary between tight and conventional, nor between tight gas and shale gas.

Coalbed Methane

Coalbed methane is the generic term given to methane gas held in coal and released or produced when the water pressure within the buried coal is reduced by pumping from either vertical or inclined to horizontal surface holes. The methane is predominantly formed during the **coalification** process whereby organic matter is slowly transformed into coal by increasing temperature and pressure as the organic matter is buried deeper and deeper by additional deposits of **organic and inorganic matter** over long periods of geological time. This is referred to as thermogenic coal bed methane. Alternatively, and more often (but not limited to) in lower rank and thermally immature coals, recent bacterial processes (involving naturally occurring bacteria associated with meteoric water recharge at **outcrop** or **subcrop**) can dominate the generation of CBM. This is referred to as late – stage biogenic coal bed methane.

During the coalification process, a range of chemical reactions takes place that produce substantial quantities of gas. While much of this gas escapes into the overlying or underlying rock, a large amount is retained within the forming coal seams. However, unlike conventional natural gas reservoirs, where gas is trapped in the **pore** or void spaces of a rock, such as sandstone, methane formed and trapped in coal is actually adsorbed onto the coal grain surfaces or **micropores** and held in place by reservoir (water) pressure. Therefore, because the micropore surface area is very large, coal can potentially hold significantly more methane per unit volume than most sandstone reservoirs.

The amount of methane stored in coal is closely related to the rank and depth of the coal; the higher the coal rank and the deeper the coal seam is presently buried (causing pressure on coal) the greater its capacity to produce and retain methane gas. Because coal has a very large internal surface area of over 1 billion square feet per ton of coal, it can hold on average three times as much gas in place as the same volume of a conventional sandstone reservoir at equal depth and pressure. In order to allow the "absorbed" gas to be released from the coal it is often necessary to lower the pressure on the coal. This generally involves removing the water contained in the coal bed. After the gas is released from the internal surfaces of the coal it moves through the internal matrix of the coal until it reaches natural fracture networks in the coal known as **cleats**. The gas then flows through these cleats or fractures until it reaches the well bore. Gas derived from coal is generally pure and requires little or no processing because it is solely methane and not mixed with heavier hydrocarbons, such as ethane, which is often present in conventional natural gas. Coal bed methane has a slightly higher energy value than some natural gases.

Natural Gas Properties

Chemical and Physical Properties

Natural gas is colorless, **odorless**, tasteless, shapeless, and lighter than air. The natural gas after appropriate treatment for acid gas reduction, odorization, and hydrocarbon and moisture dew point adjustment would then be sold within prescribed limits of pressure, **calorific value**, and possibly ***Wobbe index*** (often referred to as the *Wobbe number*). The Wobbe index (calorific value divided by the specific gravity) gives a measure of the heat input to an appliance through a given **aperture** at a given gas pressure.

Since natural gas as delivered to pipelines has practically no odor, the addition of an odorant is required by most regulations in order that the presence of the gas can be detected readily in case of accidents and leaks.

This odorization is provided by the addition of trace amounts of some organic sulfur compounds to the gas before it reaches the consumer. The sulfur compound, a chemical odorant (a ***mercaptan*** also called a *thiol* with the general formula R – SH and the odor of rotten eggs), is added to natural gas so that it can be smelled if there is a gas leak. The standard requirement is that a user will be able to detect the presence of the gas by odor when the concentration reaches 1% of gas in air. Since the lower limit of **flammability** of natural gas is approximately 5%, this requirement is equivalent to one – fifth the lower limit of flammability. The **combustion** of these trace amounts of odorant does not create any serious problems of sulfur content or toxicity.

Words and Expressions

matrix	基体
odorization	添味
coalification	成煤
outcrop	露头
subcrop	潜伏露头
pore	孔隙
micropore	微孔隙
odorless	无味
cleat	楔子;割理;夹具
aperture	孔;穴;缝隙
mercaptan	硫醇
flammability	易燃性
combustion	燃烧

Phrases and Expressions

tight gas	致密气
shale gas	页岩气
artificial stimulation	人工增产
hydraulic fracturing	水力压裂
horizontal drilling	水平钻井
organic and inorganic matter	有机质和无机质
calorific value	发热量
Wobbe index	发热量与密度的比值

Language Focus

1. Shale gas, one of the most rapidly growing forms of natural gas, is defined as natural gas from shale formations.

(参考译文:页岩气是产自页岩地层的天然气,是增长最迅速的天然气形式中的一种。)

本句中"Shale gas"是主语,"one of the most rapidly growing forms of natural gas"是主语的同位语,"be defined as"意为"定义为"。

2. Coal bed methane is the generic term given to methane gas held in coal

and released or produced when the water pressure within the buried coal is reduced by pumping from either vertical or inclined to horizontal surface holes.

（参考译文：煤层气是指储存在煤中的甲烷气。通过直井、斜井或者水平井中泵的抽汲，煤层的水压下降，煤层气释放排出。）

本句中"held in coal and released or produced when"是过去分词短语作后置定语修饰"methane gas"；"when"引导时间状语从句"the water pressure within the buried coal is reduced by pumping from either vertical or inclined to horizontal surface holes"；"either... or..."为固定搭配，意为"两者中任意一个"。

3. However, unlike conventional natural gas reservoirs, where gas is trapped in the pore or void spaces of a rock, such as sandstone, methane formed and trapped in coal is actually adsorbed onto the coal grain surfaces or micropores and held in place by reservoir (water) pressure.

（参考译文：常规气藏中气体圈闭在孔隙里，而煤层气则不同。在地层压力下，煤层气实际上是吸附在煤颗粒表面或微孔隙中。）

本句中"unlike"是介词，表示"和……不同"；"where"引导非限制性定语从句，先行词是"conventional natural gas reservoirs"，相当于"in conventional natural gas reservoirs"；句子的主语是"methane formed and trapped in coal"。

🔲 Reinforced Learning

Ⅰ. Answer the following questions for a comprehension of the text.

1. Why is the rapid rise in shale production the most significance trend in US natural gas production?

2. What is coal bed methane?

3. How is coal bed methane formed?

4. What physical properties does natural gas have?

5. What chemical properties does natural gas have?

Ⅱ. Multiple choice: choose the correct one from the alternative answers to give the exact meaning of the words.

1. In large measure this is attributable to significant advances in the use of horizontal drilling and well stimulation technologies and refinement in the cost – effectiveness of these technologies.

A. exploring   B. digging   C. exploitation   D. casting

2. During the coalification process, a range of chemical reactions takes place that produce substantial quantities of gas.

A. a series of　　B. some　　C. more　　D. few

3. Therefore, because the micropore surface area is very large, coal can potentially hold significantly more methane per unit volume than most sandstone reservoirs.

A. sound　　B. number　　C. length　　D. capacity

4. Because coal has a very large internal surface area of over 1 billion square feet per ton of coal, it can hold on average three times as much gas in place as the same volume of a conventional sandstone reservoir at equal depth and pressure.

A. largely　　B. fairly　　C. generally　　D. particularly

5. The gas then flows through these cleats or fractures until it reaches the well bore.

A. bit　　B. well　　C. oil　　D. rock

6. Gas derived from coal is generally pure and requires little or no processing because it is solely methane and not mixed with heavier hydrocarbons, such as ethane, which is often present in conventional natural gas.

A. adding　　B. operation　　C. drilling　　D. exploration

7. Natural gas is colorless, odorless, tasteless, shapeless, and lighter than air.

A. tasteless　　B. empty　　C. weightless　　D. valueless

8. The natural gas after appropriate treatment for acid gas reduction, odorization, and hydrocarbon and moisture dew point adjustment would then be sold within prescribed limits of pressure, calorific value, and possibly *Wobbe index* (often referred to as the *Wobbe number*).

A. special　　B. serious　　C. proper　　D. long – term

9. The Wobbe index (calorific value divided by the specific gravity) gives a measure of the heat input to an appliance through a given aperture at a given gas pressure.

A. rock　　B. equipment　　C. derrick　　D. drill bit

10. Since the lower limit of flammability of natural gas is approximately 5%, this requirement is equivalent to one – fifth the lower limit of flammability.

A. about　　B. nearly　　C. usually　　D. always

III. Multiple choice: read the four suggested translations and choose the best answer.

1. Shale gas, one of the most rapidly growing forms of natural gas, is defined as natural gas from shale formations.

  A. 页岩气   B. 天然气   C. 煤层气   D. 伴生气

2. The shale acts as both the source and the reservoir for the natural gas. Older shale gas wells were vertical while more recent wells are primarily horizontal and need artificial stimulation, like hydraulic fracturing, to produce.

  A. 仿造的   B. 虚伪的   C. 控制的   D. 人工的

3. Coal bed methane is the generic term given to methane gas held in coal and released or produced when the water pressure within the buried coal is reduced by pumping from either vertical or inclined to horizontal surface holes.

  A. 煤床气   B. 碳床气   C. 煤层气   D. 碳甲烷

4. The methane is predominantly formed during the coalification process whereby organic matter is slowly transformed into coal by increasing temperature and pressure as the organic matter is buried deeper and deeper by additional deposits of organic and inorganic matter over long periods of geological time.

  A. 焦煤   B. 成煤   C. 煤炭   D. 煤气

5. However, unlike conventional natural gas reservoirs, where gas is trapped in the pore or void spaces of a rock, such as sandstone, methane formed and trapped in coal is actually adsorbed onto the coal grain surfaces or micropores and held in place by reservoir (water) pressure.

  A. 岩石   B. 砂石   C. 空隙   D. 岩层

6. In order to allow the "absorbed" gas to be released from the coal it is often necessary to lower the pressure on the coal.

  A. 排出   B. 排放   C. 钻探   D. 气解

7. After the gas is released from the internal surfaces of the coal it moves through the internal matrix of the coal until it reaches natural fracture networks in the coal known as cleats.

  A. 骨架   B. 裂缝   C. 割理   D. 煤层

8. The natural gas after appropriate treatment for acid gas reduction, odorization, and hydrocarbon and moisture dew point adjustment would then be sold within prescribed limits of pressure, calorific value, and possibly *Wobbe index* (often referred to as the *Wobbe number*).

    A. 绝热指数　　　　　　　　B. 华白指数

    C. 商品指数　　　　　　　　D. 发热量与密度的比值

9. The Wobbe index (calorific value divided by the specific gravity) gives a measure of the heat input to an appliance through a given <u>aperture</u> at a given gas pressure.

    A. 管道　　　　B. 孔径　　　　C. 通道　　　　D. 管径

10. The sulfur compound, a chemical <u>odorant</u> (a *mercaptan* also called a *thiol* with the general formula R – SH and the odor of rotten eggs), is added to natural gas so that it can be smelled if there is a gas leak.

    A. 硫化物　　　B. 添味剂　　　C. 煤化气　　　D. 易燃品

IV. Put the following sentences into Chinese.

1. Older shale gas wells were vertical while more recent wells are primarily horizontal and need artificial stimulation, like hydraulic fracturing, to produce.

2. In large measure this is attributable to significant advances in the use of horizontal drilling and well stimulation technologies and refinement in the cost – effectiveness of these technologies.

3. The amount of methane stored in coal is closely related to the rank and depth of the coal; the higher the coal rank and the deeper the coal seam is presently buried (causing pressure on coal) the greater its capacity to produce and retain methane gas.

4. Because coal has a very large internal surface area of over 1 billion square feet per ton of coal, it can hold on average three times as much gas in place as the same volume of a conventional sandstone reservoir at equal depth and pressure.

5. The natural gas after appropriate treatment for acid gas reduction, odorization, and hydrocarbon and moisture dew point adjustment would then be sold within prescribed limits of pressure, calorific value, and possibly *Wobbe index.*

V. Put the following paragraphs into Chinese.

1. Tight gas is a general term for natural gas found in low permeability formations. Generally, we classify as tight gas those low permeability gas reservoirs that cannot produce economically without the use of technologies to stimulate flow of the gas towards the well, such as hydraulic fracturing. Tight gas is often a poorly defined category with no clear boundary between tight and convention-

al, nor between tight gas and shale gas.

2. Since natural gas as delivered to pipelines has practically no odor, the addition of an odorant is required by most regulations in order that the presence of the gas can be detected readily in case of accidents and leaks. This odorization is provided by the addition of trace amounts of some organic sulfur compounds to the gas before it reaches the consumer. The sulfur compound, a chemical odorant, is added to natural gas so that it can be smelled if there is a gas leak.

3.3　Gas Transportation (1)

⊞ Guidance to Reading

*Natural gas extracted from wells is called casinghead gas, which has to be carried to the field processing plant for treatment and then to consumers. Because of its low density, it is not easy to store natural gas or transport by vehicle. Gas pipeline systems are remarkable for their efficiency and low transportation cost, however, not practical across oceans. LNG carriers are used to transport liquefied natural gas (LNG) across oceans, while tank trucks can carry liquefied or **compressed natural gas** (CNG) over shorter distances. Sea transport using CNG carrier ships that are now under development may be competitive with LNG transport in specific conditions.*

⊞ Text

Gas, as a result of the storage difficulties, needs to be transported immediately to its destination after production from a reservoir. There are a number of options for transporting natural gas energy from oil and gas fields to market. These include pipelines, liquefied natural gas (LNG), compressed natural gas (CNG), and so on.

Pipelines

Pipelines are a very convenient method of transport but are not flexible as the gas will leave the source and arrive at its (one) destination. If the pipeline has to be shut down, the production and receiving facilities and **refinery** often also have to be shut down because gas cannot be readily stored, except perhaps by increasing the pipeline pressure by some percentage.

In the last decade, on average, over 12,000 miles per year of new gas pipelines have been completed; most are transnational. If political stability can be guaranteed, pipelines may be able to provide a long – term solution for transportation. An example of this approach is a proposed deep water pipeline from Oman to India. However, the cost of building such a pipeline remains unclear. Subsea lines over 2000 miles have, until recently, been regarded as uneconomic because of the subsea terrain making pipeline installation and **maintenance** expensive and any recompression along the route difficult, but changes are in the air! If technical and economic hurdles can be overcome, these pipelines can become effective.

Liquefied Natural Gas

LNG is the liquid form of natural gas. Gas cooled to approximately $-162^\circ C$ liquefies and has a volume approximately 1/600 that of gas at room temperature. However, facilities for liquefying natural gas require complex machinery with moving parts and special refrigerate ships for transporting the liquefied natural gas to market. The costs of building a liquefied natural gas plant have lowered since the mid – 1980s because of greatly improved **thermodynamic** efficiencies, making liquefied natural gas a major gas export method worldwide, and many plants are being extended or new ones are being built in the world.

Large **cryogenic** tanks are needed to store the liquefied natural gas; typically these may be 70 m in diameter, 45 m high, and hold over $100,000$ m^3 of liquefied natural gas. At the consumer end, an **infrastructure** for handling the reprocessing of vast quantities of natural gas from LNG is required, which is also expensive and **vulnerable** to **sabotage**.

The current largest specially built refrigerated tankers can carry $135,000$ m^3 of liquefied natural gas, **equivalent to** 2.86 billion scf of gas, but are very expensive. This makes it difficult for liquefied natural gas to use smaller isolated (offshore) reserves and to serve small markets commercially because it is this large capacity and continuous running that keep thermodynamic efficiency high and costs to a minimum. Thus small volumes of **intermittent** gas are not economically attractive to the major gas sellers for liquefied natural gas facilities. However, a small **well – insulated** liquefied natural gas container trade is being investigated, and, if successful, small quantities of liquefied natural gas may be able to be delivered from liquefied natural gas storage, just like the **gas-**

oline tankers of today. Even so, liquefied natural gas must be stored for periods of time (months) without significant boil – off losses, which is difficult.

Compressed Natural Gas

Gas can be transported in containers at high pressures, typically 1800 psig for a rich gas (significant amounts of ethane, propane, etc.) to roughly 3600 psig for a lean gas (mainly methane). Gas at these pressures is termed *compressed natural gas*. Compressed natural gas is used in some countries for **vehicular** transport as an **alternative** to conventional fuels (gasoline or **diesel**). The filling stations can be supplied by pipeline gas, but the compressors needed to get the gas to 3000 psig can be expensive to purchase, maintain, and operate.

An alternative approach has dedicated transport ships carrying straight long, large – diameter pipes in an insulated cold storage cargo package. The gas has to be dried, compressed, and chilled for storage onboard. By careful control of temperature, more gas should be transported in any ship of a given payload capacity, subject to volume limitation and amount and weight of material of the pipe (pressure and safety considerations). Suitable compressors and **chillers** are needed, but would be much less expensive than a natural gas liquefier and would be standard so that costs could be further minimized. According to the **proponents**, the terminal facilities would also be simple and hence less expensive. Two new types of CNG transport are being promoted by their respective companies and are discussed next :

"VOTRANS" is a new type of CNG marine – transport technology from EnerSea Transport, L. L. C. Its engineering studies indicate that it can move up to 2 Bcf per ship over distances up to 4000 miles at significantly lower total costs than LNG. The technology comprises large – diameter pipe structures manifolded together in tiers, essentially a sea – going pipeline.

To maintain temperature, the pipe structures are contained within a nitrogen – filled, insulated chamber. It can store CNG more efficiently at significantly lower compression (40% compared to LNG), increase vessel capacities, reduce costs, and transport both lean and rich gas. Finally, VOTRANS minimizes gas losses during processing and transport to less than 7% compared to as much as 20% for LNG.

"Coselle" CNG technology is from Cran & Stenning Technology Inc. The system uses conventional, 10. 6 – mile – long, 6 – in diameter, 1/4 – in wall thickness line pipe in large coils (coselles). Such a CNG carrier may have 108 coselles with a 330 – MMcfg capacity. Stored gas temperature is 50 °F at 3000 psi. American Bureau of Shipping and Det Norske Veritas have concluded that a Coselle CNG carrier is "at least as safe as other gas carriers. " These ships can be loaded at relatively simple marine facilities, including offshore buoy moorings, through flexible hoses connected to onshore or on – platform compressor stations. "Coselle" and "VOTRANS" are two would – be commercial, highpressure gas storage and transport technologies for CNG.

Compressed natural gas technology provides an effective way for shorter – distance transport of gas. The technology is aimed at monetizing offshore reserves, which cannot be produced because of the **unavailability** of a pipeline or because the LNG option is very costly. Technically, CNG is easy to deploy with lower requirements for facilities and infrastructure.

Results show that for distances up to 2500 miles, natural gas can be transported as CNG at prices ranging from ＄ 0. 93 to ＄ 2. 23 per MMBTU compared to LNG, which can cost anywhere from ＄ 1. 5 to ＄ 2. 5 per MMBTU depending on the actual distance. At distances above 2500 miles the cost of delivering gas as CNG becomes higher than the cost for LNG because of the disparity in the volumes of gas transported with the two technologies.

Words and Expressions

refinery	处理厂
maintenance	维护
thermodynamic	热动力学
cryogenic	冷冻的
infrastructure	基础设施
vulnerable	易受伤害的
sabotage	破坏
intermittent	间歇的
gasoline	汽油
vehicular	车辆的
alternative	交替的
diesel	柴油
chiller	冷却机

| proponent | 支持者 |
| unavailability | 局限 |

Phrases and Expressions

compressed natural gas	压缩天然气
equivalent to	相当于
well – insulated	密封隔热性能良好

Language Focus

1. If the pipeline has to be shut down, the production and receiving facilities and refinery often also have to be shut down because gas cannot be readily stored, except perhaps by increasing the pipeline pressure by some percentage.

（参考译文：由于气体不容易储存，除非适当地增加管线压力，否则管线出现故障，气井、集气站、处理厂都得停工。）

本句中"if"引导条件状语从句"the pipeline has to be shut down"；"because"引导原因状语从句"gas cannot be readily stored, except perhaps by increasing the pipeline pressure by some percentage"，"except"作介词，表示"除……之外"。

2. Subsea lines over 2000 miles have, until recently, been regarded as uneconomic because of the subsea terrain making pipeline installation and maintenance expensive and any recompression along the route difficult, but changes are in the air!

（参考译文：目前认为修建超过2000mi的海底管线不经济，因为海底管线的敷设和维护费用相当高昂，而且沿管线进行再压缩非常困难，至今还没有很好的解决方法。）

本句的主谓结构是"Subsea lines over 2000 miles have been regarded as..."；"until recently"是插入语；"because of the subsea terrain making pipeline installation and maintenance expensive and any along the route difficult"是原因状语；"but"是转折连词，引导另一个分句"changes are in the air"。

3. The costs of building a liquefied natural gas plant have lowered since the mid – 1980s because of greatly improved thermodynamic efficiencies, making liquefied natural gas a major gas export method worldwide, and many plants are being extended or new ones are being built in the world.

（参考译文：自20世纪80年代中期以来，由于热动力学的发展，液化天然气工厂的建造成本下降，这使得LNG技术成为备受世界各国青睐的天然气输送方式。现在，许多LNG工厂扩张规模，更多的工厂正在建设。）

本句由三个并列句构成,分别由"and"和"or"连接。"since the mid - 1980s"是时间状语,表示"自 20 世纪 80 年代中期以来";"because of greatly improved thermodynamic efficiencies"是原因状语;"making liquefied natural gas a major gas export method worldwide"是伴随状语。

4. This makes it difficult for liquefied natural gas to use smaller isolated (offshore) reserves and to serve small markets commercially because it is this large capacity and continuous running that keep thermodynamic efficiency high and costs to a minimum.

(参考译文:LNG 技术是基于大批量输送和连续运行来提高效率从而降低成本的,因此对于孤立的储层(如海上)或者需求较小的用户,LNG 技术的推广应用难度较大。)

本句中"because"引导原因状语从句"it is this large capacity and continuous running that keep thermodynamic efficiency high and costs to a minimum",其中"It is... that"为强调句型,强调的内容放在"it is"后面和"that"前面。

5. Results show that for distances up to 2500 miles, natural gas can be transported as CNG at prices ranging from $ 0. 93 to $ 2. 23 per MMBTU compared to LNG, which can cost anywhere from $ 1. 5 to $ 2. 5 per MMB-TU depending on the actual distance.

(参考译文:结果表明,在 2500mi 范围内 LNG 输送天然气的成本为 1.5 ~2.5 美元/10^6Btu,而 CNG 输送成本为 0.93 ~2.23 美元/10^6Btu。)

本句中"results"是主语,"show"是谓语动词,"that"引导宾语从句"for distances up to 2500 miles, natural gas can be transported as CNG at prices ranging from $ 0. 93 to $ 2. 23 per MMBTU compared to LNG, which can cost anywhere from $ 1. 5 to $ 2. 5 per MMBTU depending on the actual distance",其中"which"引导非限制性定语从句,指代其先行词"LNG"。

Reinforced Learning

I . Answer the following questions for a comprehension of the text.

1. What is the most cost - effective way for natural gas transportation?

2. What is the role of LNG?

3. What is compressed natural gas?

4. How to transport compressed natural gas?

5. What is "Coselle" CNG technology?

II . Multiple choice : choose the correct one from the alternative answers to give the exact meaning of the words.

1. There are a number of options for transporting natural gas energy from oil and gas fields to market.

   A. ways   B. methods   C. channels   D. choices

2. If the pipeline has to be shut down, the production and receiving facilities and refinery often also have to be shut down because gas cannot be readily stored, except perhaps by increasing the pipeline pressure by some percentage.

   A. operation   B. processing factory

   C. production   D. package

3. An example of this approach is a proposed deep water pipeline from Oman to India.

   A. way   B. case   C. matter   D. event

4. Subsea lines over 2000 miles have, until recently, been regarded as uneconomic because of the subsea terrain making pipeline installation and maintenance expensive and any along the route difficult, but changes are in the air!

   A. cleaning   B. transportation   C. conveying   D. preservation

5. The costs of building a liquefied natural gas plant have lowered since the mid – 1980s because of greatly improved thermodynamic efficiencies, making liquefied natural gas a major gas export method worldwide, and many plants are being extended or new ones are being built in the world.

   A. established   B. lengthened   C. expanded   D. broadened

6. At the consumer end, an infrastructure for handling the reprocessing of vast quantities of natural gas from LNG is required, which is also expensive and vulnerable to sabotage.

   A. destruction   B. disaster   C. accident   D. explosion

7. The current largest specially built refrigerated tankers can carry 135,000 m^3 of liquefied natural gas, equivalent to 2. 86 billion scf of gas, but are very expensive.

   A. about   B. equal to   C. reaching   D. amount to

8. Compressed natural gas is used in some countries for vehicular transport as an alternative to conventional fuels.

   A. alternated   B. optional   C. selective   D. another

9. According to the proponents, the terminal facilities would also be simple

and hence less expensive.

    A. proposals   B. policies   C. principles   D. supporters

10. These ships can be loaded at relatively simple marine facilities, including offshore buoy moorings, through flexible hoses connected to onshore or on – platform compressor stations.

    A. inland factories   B. sea facilities

    C. overseas fields   D. online systems

Ⅲ. Multiple choice: read the four suggested translations and choose the best answer.

1. These include pipelines, liquefied natural gas (LNG), compressed natural gas (CNG), gas to solids (GTS), i. e. , hydrates, gas to power (GTP), i. e. , electricity, and gas to liquids (GTL), with a wide range of possible products, including clean fuels, plastic precursors, or methanol and gas to commodity (GTC), such as aluminum, glass, cement, or iron.

    A. 银、玻璃、甲烷或铁   B. 钢、玻璃、塑料或铁

    C. 铁、玻璃、钢板或器皿   D. 铝、玻璃、水泥或铁

2. Subsea lines over 2000 miles have, until recently, been regarded as uneconomic because of the subsea terrain making pipeline installation and maintenance expensive and any along the route difficult, but changes are in the air!

    A. 在空中   B. 在空气中   C. 变幻莫测   D. 悬而未决

3. The costs of building a liquefied natural gas plant have lowered since the mid – 1980s because of greatly improved thermodynamic efficiencies, making liquefied natural gas a major gas export method worldwide, and many plants are being extended or new ones are being built in the world.

    A. 蒸汽机   B. 机械   C. 热动力学   D. 能量转换

4. Large cryogenic tanks are needed to store the liquefied natural gas; typically these may be 70 m in diameter, 45 m high, and hold over $100,000$ m^3 of liquefied natural gas.

    A. 天然气瓶 B. 低温储罐   C. 液化气罐   D. 储气箱

5. Thus small volumes of intermittent gas are not economically attractive to the major gas sellers for liquefied natural gas facilities.

    A. 中断供气 B. 慢速供气   C. 间歇供气   D. 不停供气

6. However, a small well – insulated liquefied natural gas container trade is being investigated, and, if successful, small quantities of liquefied natural gas may be able to be delivered from liquefied natural gas storage, just like the gas-

oline tankers of today.

    A. 密封隔热性能良好   B. 传送速度良好

    C. 安全稳定性高   D. 质量高

    7. Compressed natural gas is used in some countries for vehicular transport as an alternative to conventional fuels (gasoline or diesel).

    A. 汽油   B. 煤油   C. 柴油   D. 鲸油

    8. An alternative approach has dedicated transport ships carrying straight long, large – diameter pipes in an insulated cold storage cargo package.

    A. 绝缘冷冻室   B. 密封隔热冷藏储罐

    C. 密封货物箱   D. 密封冷冻储藏柜

    9. By careful control of temperature, more gas should be transported in any ship of a given payload capacity, subject to volume limitation and amount and weight of material of the pipe (pressure and safety considerations).

    A. 油气运输 B. 油气管道   C. 资产能力   D. 负载能力

    10. It can store CNG more efficiently at significantly lower compression (40% compared to LNG), increase vessel capacities, reduce costs, and transport both lean and rich gas.

    A. 增加能力, 减少花费, 运输贫瘦气

    B. 减少储量, 降低成本, 储存贫瘦气

    C. 储量更大, 成本更低, 贫气富气都适用

    D. 增加储量, 增加成本, 储存贫气富气

Ⅳ. Put the following sentences into Chinese.

    1. These ships can be loaded at relatively simple marine facilities, including offshore buoy moorings, through flexible hoses connected to onshore or on – platform compressor stations.

    2. If the pipeline has to be shut down, the production and receiving facilities and refinery often also have to be shut down because gas cannot be readily stored, except perhaps by increasing the pipeline pressure by some percentage.

    3. Subsea lines over 2000 miles have, until recently, been regarded as uneconomic because of the subsea terrain making pipeline installation and maintenance expensive and any along the route difficult, but changes are in the air!

    4. The costs of building a liquefied natural gas plant have lowered since the mid – 1980s because of greatly improved thermodynamic efficiencies, making liquefied natural gas a major gas export method worldwide, and many plants are being extended or new ones are being built in the world.

5. Compressed natural gas technology is aimed at monetizing offshore reserves, which cannot be produced because of the unavailability of a pipeline or because the LNG option is very costly.

V. Put the following paragraphs into Chinese.

1. This makes it difficult for liquefied natural gas to use smaller isolated (offshore) reserves and to serve small markets commercially because it is this large capacity and continuous running that keep thermodynamic efficiency high and costs to a minimum. Thus small volumes of intermittent gas are not economically attractive to the major gas sellers for liquefied natural gas facilities. However, a small well – insulated liquefied natural gas container trade is being investigated, and, if successful, small quantities of liquefied natural gas may be able to be delivered from liquefied natural gas storage, just like the gasoline tankers of today.

2. An alternative approach has dedicated transport ships carrying straight long, large – diameter pipes in an insulated cold storage cargo package. The gas has to be dried, compressed, and chilled for storage onboard. By careful control of temperature, more gas should be transported in any ship of a given payload capacity, subject to volume limitation and amount and weight of material of the pipe (pressure and safety considerations). Suitable compressors and chillers are needed, but would be much less expensive than a natural gas liquefier and would be standard so that costs could be further minimized.

3.4 Gas Transportation (2)

🔲 Guidance to Reading

*Natural gas can be transported in many other ways besides pipelines, liquefied natural gas (LNG), and compressed natural gas (CNG). Gas can be transported as a solid in the form of gas hydrates, gas to solids (GTS); gas can be used to generate electricity and then transported to consumers, gas to power (GTP); gas can be transformed into a wide range of liquid products, including clean fuels, plastic **precursors** or methanol, gas to liquids (GTL); and gas can be transformed into commodity such as aluminum, glass, cement, or iron, gas to commodity (GTC).*

Text

Gas to Solid

Gas can be transported as a solid, with the solid being gas hydrate. **Natural gas hydrate** is the product of mixing natural gas with liquid water to form a stable water **crystalline** ice – like substance. NGH transport, which is still in the experimental stage, is believed to be a **viable** alternative to liquefied natural gas or pipelines for the transportation of natural gas from source to demand.

Gas to solids involves three stages: production, transportation, and regasification. Natural gas hydrates are created when certain small molecules, particularly methane, ethane, and propane, stabilize the hydrogen bonds within water to form a three – dimensional, cage – like structure with the gas molecule trapped within the cages. A cage is made up of several water molecules held together by hydrogen bonds. Hydrates are formed from natural gas in the presence of liquid water, provided the pressure is above and the temperature is below the equilibrium line of the phase diagram of the gas and liquid water. The solid has a snow – like appearance.

In the oil/gas industry, natural hydrates are a pipeline **nuisance** and safety hazard. Considerable care must be taken by the operators to ensure that these hydrates do not form, as they can block pipelines if precautions, such as methanol injection, are not taken. However, vast quantities of gas hydrate have been found in permafrost and at the seabed in depths below 500 m (1500 ft) and, if exploited properly, could become the major energy source in the next 30 years.

For gas transport, natural gas hydrates can be formed deliberately by mixing natural gas and water at 80 to 100 bar and 2 to 10℃. If the **slurry** is refrigerated to around −15℃, it decomposes very slowly at atmospheric pressure so that the hydrate can be transported by ship to market in simple containers insulated to near – **adiabatic** conditions. At the market, the slurry is melted back to gas and water by controlled warming for use after appropriate drying in electricity power generation stations or other requirements. The hydrate mixture yields up to 160 m^3 of natural gas per ton of hydrate, depending on the manufacture process. The manufacture of the hydrate could be carried out using mobile equipment for onshore and ship for offshore using a floating production, storage, and off – loading vessel with minimal gas processing (cleaning, etc.) prior to hydrate formation, which is attractive commercially.

Conceptually, hydrate slurry production is simply mixing chilled water and gas. In practice, processed gas is fed to a hydrate production plant, where a series of reactors convert it into hydrate slurry. Each reactor further concentrates the hydrate slurry. It is then stored and eventually offloaded onto a transport vessel. At the receiving terminal, the hydrate is dissociated and the gas can be used as desired. The water can be used at the destination if there is a water shortage or returned as **ballast** to the hydrate generator; because it is saturated with gas, the water will not take more gas into solution.

The hydrate mixture can be stored at normal temperatures (0 to $-10\,℃$) and pressures (10 to 1 atmosphere) where 1 m^3 of hydrate should contain about 160 m^3 gas. This concentration of gas is attractive, as it is easier to produce, safer, and less expensive to store compared to the 200 m^3 gas per 1 m^3 of compressed gas (high pressure ca. 3000 psig) or the 637 m^3 gas per 1 m^3 of liquefied natural gas (low temperatures of $-162\,℃$).

Gas storage in hydrate form becomes especially efficient at relatively low pressures where substantially more gas **per unit volume** is contained in the hydrate than in the free state or in the compressed state when the pressure has dropped. When compared to the transportation of natural gas by pipeline or as liquefied natural gas, the hydrate concept has lower capital and operating costs for the movement of quantities of natural gas over **adverse conditions**. Thus, gas hydrate is very effective for gas storage and transport as it eliminates low temperatures and the necessity of compressing the gas to high pressures. Dry hydrate pellets yield about 160 m^3 of gas at standard conditions from 1 m^3 of hydrate compared to the 637 m^3 per 1 m^3 of liquefied natural gas. This is a considerable volume penalty (and hence transport cost) if considered **in isolation**; with less expensive ships for hydrate transport, the process could be economic.

Gas to Power

Currently, much of the transported gas destination is fuel for electricity generation. Electricity generation at or near the reservoir source and transportation by cable to the destination(s) (GTP) is possible. Thus, for instance, offshore or isolated gas could be used to fuel an offshore power plant (may be sited in **less hostile waters**) , which would generate electricity for sale onshore or to other offshore customers. Unfortunately, because installing high – power lines to reach the shoreline appears to be almost as expensive as pipelines, that gas to power could be viewed as defeating the purpose of an alternative less expensive solution for transporting gas.

There is significant energy loss from the cables along the long – distance transmission lines, more so if the power is AC rather than DC; additionally, losses also occur when the power is converted to DC from AC and when it is converted from the **high voltages** used in transmission to the lower values needed by the consumers.

There are other practical considerations to note such as if the gas is associated gas, then if there is a generator shutdown and no other gas outlet, the whole oil production facility might also have to be shut down or the gas released to flare. Also, if there are operational problems within the generation plant the generators must be able to shut down quickly (in around 60 s) to keep a small incident from escalating. Additionally, the shutdown system itself must be safe so that any plant that has complicated processes that require a **purge cycle** or a cool – down cycle before it can shut down is clearly unsuitable. Finally, if the plant cannot shut down easily and/or be able to start up again quickly (perhaps in an hour), operators will be hesitant to ever shut down the process for fear of financial **retribution** from the power distributors.

Gas to Liquids

In GTL transport processes, the natural gas is converted to a liquid, such as **syncrude** methanol and **ammonia**, and is transported as such. The technology of converting natural gas to liquids is not new. In the first step, methane is mixed with steam and converted to **syngas** or synthetic gas (mixtures of carbon monoxide and hydrogen) by one of a number of routes using suitable new **catalyst** technology. The syngas is then converted into a liquid. The produced liquid can be a fuel, usually a clean – burning motor fuel (syncrude) or **lubricant**, or ammonia or methanol or some precursor for plastics manufacture. Hundreds of modifications and patents have been applied to this complex, energy – intensive process. Most recent modifications generally involve lowering capital expenditures and the overall energy required for processing, especially through the use of proprietary catalysts and the manner in which oxygen is added to the system.

Words and Expressions

precursor   先驱;前体

crystalline	透明的
viable	可行的
nuisance	损害
slurry	泥浆
adiabatic	绝热的
ballast	压舱物
retribution	报应
syncrude	合成原油
ammonia	氨
syngas	合成气
catalyst	催化剂
lubricant	润滑剂

Phrases and Expressions

natural gas hydrate	天然气水合物
per unit volume	单位体积
adverse condition	不利条件
in isolation	孤立地
less hostile waters	险恶水域
high voltage	高电压
purge cycle	净化处理

Language Focus

1. Hydrates are formed from natural gas in the presence of liquid water, provided the pressure is above and the temperature is below the equilibrium line of the phase diagram of the gas and liquid water.

(参考译文:当压力高于气水两相相图中的平衡线而温度低于此平衡线时,天然气与水形成雪花状的水合物。)

本句中"provided"作连词引导条件状语从句"the pressure is above and the temperature is below the equilibrium line of the phase diagram of the gas and liquid water",意为"假如,倘若"。

2. Considerable care must be taken by the operators to ensure that these hydrates do not form, as they can block pipelines if precautions, such as methanol injection, are not taken.

（参考译文：在输送过程中必须非常小心，防止水合物生成。如果未采取预防措施便注入甲醇，管线就可能堵塞。）

本句中主谓结构为被动语态，"to ensure that these hydrates do not form"为目的状语，"that"引导宾语从句"these hydrates do not form"；"as"引导原因状语从句"they can block pipelines if precautions, such as methanol injection, are not taken"，"such as methanol injection"为插入语，放在"if"条件句主语"precautions"与谓语"are not taken"中间。

3. The produced liquid can be a fuel, usually a clean – burning motor fuel (syncrude) or lubricant, or ammonia or methanol or some precursor for plastics manufacture.

（参考译文：产出液可以是燃料，通常是洁净的动力燃料（合成石油）或润滑剂、氨、甲醇、制造塑料的原料。）

本句主谓结构是"The produced liquid can be a fuel"，后面"usually a clean – burning motor fuel (syncrude) or lubricant, or ammonia or methanol or some precursor for plastics manufacture"是"a fuel"的同位语，"or"连接并列的名词短语。

Reinforced Learning

I. Answer the following questions for a comprehension of the text.

1. What is natural gas hydrate?

2. How are natural gas hydrates created?

3. What condition is necessary for gas hydrates to be transported?

4. What is the bottleneck for gas – to – power natural gas transportation?

5. What can GTL be used for?

II. Multiple choice: choose the correct one from the alternative answers to give the exact meaning of the words.

1. Natural gas hydrate is the product of mixing natural gas with liquid water to form a stable water crystalline ice – like substance.

　　A. rock　　　B. white　　　C. clear　　　D. bright

2. NGH transport, which is still in the experimental stage, is believed to be a viable alternative to liquefied natural gas or pipelines for the transportation of natural gas from source to demand.

　　A. livable　　B. practical　　C. various　　D. wonderful

3. Hydrates are formed from natural gas in the presence of liquid water, provided the pressure is above and the temperature is below the <u>equilibrium</u> line of the phase diagram of the gas and liquid water.

    A. straight   B. balance   C. rising   D. curve

4. In the oil/gas industry, natural hydrates are a pipeline <u>nuisance</u> and safety hazard.

    A. problem   B. channel   C. harm   D. bond

5. When compared to the transportation of natural gas by pipeline or as liquefied natural gas, the hydrate concept has lower capital and operating costs for the movement of quantities of natural gas over <u>adverse</u> conditions.

    A. bad   B. fatal   C. beneficial   D. advantageous

6. Some consider having the energy as gas at the consumers' end gives greater flexibility and better thermal efficiencies because the waste heat can be used for local heating and <u>desalination</u>.

    A. electricity   B. production

    C. accommodation   D. salt removal

7. Also, if there are operational problems within the generation plant the generators must be able to shut down quickly (in around 60 s) to keep a small incident from <u>escalating</u>.

    A. growing   B. changing   C. disappearing   D. solving

8. <u>Nevertheless</u>, gas to power has been an option much considered in the United States for getting energy from the Alaskan gas and oil fields to populated areas.

    A. Although   B. If   C. Whereas   D. Besides

9. Additionally, the shutdown system itself must be safe so that any plant that has complicated processes that require <u>a purge cycle</u> or a cool – down cycle before it can shut down is clearly unsuitable.

    A. cooling   B. warming   C. a trial   D. cleaning

10. Hundreds of <u>modifications</u> and patents have been applied to this complex, energy – intensive process, and further developments continue to the present day.

    A. failure   B. improvements

    C. experiments   D. creation

Ⅲ. Multiple choice：read the four suggested translations and choose the best answer.

1. Gas to solids involves three stages：production，transportation，and regasification.

A. 重新生气　　B. 重新变气　　C. 再汽化　　D. 再产气

2. The water can be used at the destination if there is a water shortage or returned as ballast to the hydrate generator；because it is saturated with gas，will not take more gas into solution.

A. 储气罐　　B. 干燥剂　　C. 饱和物　　D. 压舱物

3. There is significant energy loss from the cables along the long – distance transmission lines，more so if the power is AC rather than DC；additionally，losses also occur when the power is converted to DC from AC and when it is converted from the high voltages used in transmission to the lower values needed by the consumers.

A. 交流电　直流电　　　　　　　B. 直流电　交流电

C. 并行　串联　　　　　　　　　D. 串联　并行

4. This view is strengthened by economics，as power generation uses approximately 1 million scf/day of gas for every 10 MW of power generated so that even large generation capacity would not consume much of the gas from larger fields and thus not generate large revenues for the gas producers.

A. 效益　　B. 利润　　C. 财富　　D. 利益

5. Finally，if the plant cannot shut down easily and/or be able to start up again quickly（perhaps in an hour），operators will be hesitant to ever shut down the process for fear of financial retribution from the power distributors.

A. 金融收益　　B. 金融回报　　C. 经济报复　　D. 经济索赔

6. In GTL transport processes，the natural gas is converted to a liquid，such as syncrude methanol and ammonia，and is transported as such.

A. 氨、甲醇和尿素　　　　　　　B. 甲烷、乙烷和丙烷

C. 合成原油、甲醇和氨　　　　　D. 铝、玻璃和水泥

7. In the first step，methane is mixed with steam and converted to syngas or synthetic gas（mixtures of carbon monoxide and hydrogen）by one of a number of routes using suitable new catalyst technology.

A. 合成气　　B. 合成原油　　C. 催化剂　　D. 动力燃料

8. The syngas is then converted into a liquid using a Fischer – Tropsch

process (in the presence of a catalyst) or an <u>oxygenation</u> method (mixing syngas with oxygen in the presence of a suitable catalyst).

    A. 氧气   B. 氧气混合物 C. 加氧法   D. 氧气合成

9. The produced liquid can be a fuel, usually a clean – burning motor fuel (syncrude) or lubricant, or <u>ammonia</u> or methanol or some precursor for plastics manufacture.

    A. 尿素   B. 二乙醚   C. 甲醇   D. 氨

10. Methanol can be used in <u>internal combustion engines</u> as a fuel, but the current market for methanol as a fuel is limited, although the development of fuel cells for motor vehicles may change this.

    A. 内燃机   B. 内引擎   C. 内缸   D. 燃烧动力

IV. Put the following sentences into Chinese.

1. NGH transport, which is still in the experimental stage, is believed to be a viable alternative to liquefied natural gas or pipelines for the transportation of natural gas from source to demand.

2. In the oil/gas industry, natural hydrates are a pipeline nuisance and safety hazard.

3. Considerable care must be taken by the operators to ensure that these hydrates do not form, as they can block pipelines if precautions, such as methanol injection, are not taken.

4. However, vast quantities of gas hydrate have been found in permafrost and at the seabed in depths below 500 m and, if exploited properly, could become the major energy source in the next 30 years.

5. Other GTL processes are being developed to produce clean fuels, e. g. , syncrude, diesel, or many other products, including lubricants and waxes, from gas but require a complex (expensive) chemical plant with novel catalyst technology.

V. Put the following paragraphs into Chinese.

1. Because installing high – power lines to reach the shoreline appears to be almost as expensive as pipelines, that gas to power could be viewed as defeating the purpose of an alternative less expensive solution for transporting gas. In addition, there is significant energy loss from the cables along the long – distance transmission lines, more so if the power is AC rather than DC; additionally, los-

ses also occur when the power is converted to DC from AC and when it is converted from the high voltages used in transmission to the lower values needed by the consumers.

2. If the gas used for electricity generation is associated gas, then if there is a generator shutdown and no other gas outlet, the whole oil production facility might also have to be shut down or the gas released to flare. Also, if there are operational problems within the generation plant the generators must be able to shut down quickly to keep a small incident from escalating. Additionally, the shutdown system itself must be safe so that any plant that has complicated processes that require a purge cycle or a cool – down cycle before it can shut down is clearly unsuitable. Finally, if the plant cannot shut down easily and/or be able to start up again quickly, operators will be hesitant to ever shut down the process for fear of financial retribution from the power distributors.

Chapter 4　Natural Gas Separation

4.1　The Basics of Natural Gas Processing

🔲 Guidance to Reading

*Although the processing of natural gas is in many respects less complicated than the processing of crude oil, it is equally as necessary before its use by end users. The natural gas used by consumers is composed almost entirely of methane. Whatever the source of the natural gas (from oil, gas or **condensate** reservoir), it commonly exists in mixtures with other hydrocarbon, principally ethane, propane, butane, and pentanes. In addition, raw natural gas contains water vapor, hydrogen sulfide (H_2S), carbon dioxide, helium, nitrogen, and other compounds. Natural gas processing is to produce what is known as "pipeline quality" dry natural gas.*

🔲 Text

Raw natural gas after transmission through the field – gathering network must be processed before it can be moved into long – distance pipeline systems for use by consumers. The objective of gas processing is to separate natural gas, condensate, **noncondensable**, **acid gases**, and water from a gas – producing well and condition these fluids for sale or disposal. The typical process operation modules are shown in Figure 4. 1. 1 Each module consists of a single piece or a group of equipment performing a specific function. All the modules shown will not necessarily be present in every gas plant. In some cases, little processing is needed; however, most natural gas requires processing equipment at the gas processing plant to remove **impurities**, water, and excess hydrocarbon liquid and to control delivery pressure. The unit operations used in a given application may not be arranged in the sequence shown in Figure 4. 1. 1, although this sequence is typical. The choice of modules to be used and the arrangement of these modules are determined during the design stage of each gas – field development project.

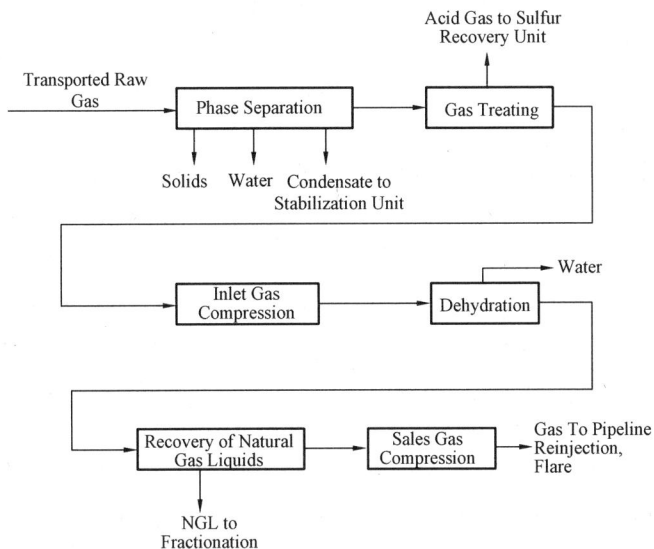

Figure 4.1.1 Simplified typical onshore treatment process

Process Modules

The first unit module is the physical separation of the distinct phases, which are typically gas, liquid hydrocarbons, liquid water, and/or solids. Phase separation of the production stream is usually performed in an inlet separator. Inlet gas receiving is complicated by the fact that transmission lines supplying the plant typically operate with two or three phases present and consequently liquid slugging is common. Slugs are normally formed from **elevational changes** in the inlet supply pipes, changes in gas supply flow rates, and changes in pressure and temperature during transmission. Slug flow may even be encountered in horizontal pipes under steady – state conditions if the flow regime is not properly selected. The arrival of "slugs" at production or processing equipment impacts the operation of production facilities negatively, causing both mechanical problems (due to **high velocities and momentum**) and process problems (increasing liquid levels, causing **surges** and **trips**).

Gas pipelines have typically used **slug catchers** to **dissipate** the energy of the liquid slugs, to minimize **turbulence**, to ensure that the gas and liquid flow rates are low enough so that the **stratified flow** regime and subsequently **gravity segregation** can occur. The slug catcher is designed to separate gas, hydrocarbon condensate, and inlet water. The liquids that collect in the slug catcher flow to a three – phase separator from which the two liquid phases, hydrocarbon conden-sate and water/methanol or water/glycol phases, are outputs. Overhead gas from the three – phase separator is recompressed where necessary for use as fuel gas.

Hydrocarbon condensate recovered from natural gas may be shipped with-out further processing but is typically stabilized to produce a safe transportable liquid. Unstabilized condensates contain a large percentage of methane and eth-ane, which will vaporize easily in storage tanks. Stabilization is the full removal of light fractions from the condensate, usually achieved by distillation. Stabilized liquid will generally have a vapor pressure specification (**Reid vapor pressure** of < 10 psi), as the product will be injected into a pipeline or transport pressure vessel, which has definite pressure limitations.

The next step in natural gas processing is acid gas treating. In addition to heavy hydrocarbons and water vapor, natural gas often contains other contami-nants that may have to be removed. Carbon dioxide (CO_2), hydrogen sulfide (H_2S), and other sulfur – containing species such as **mercaptans** are com-pounds that require complete or partial removal. These compounds are collec-tively known as "acid gases. " H_2S when combined with water forms a weak sulfuric acid, whereas CO_2 and water form **carbonic acid**, thus the term "acid gas. " Natural gas with H_2S or other sulfur compounds present is called "sour gas," whereas gas with only CO_2 is called "sweet. " Both H_2S and CO_2 are very undesirable, as they cause corrosion and present a major safety risk.

Depending on the pressure at the plant gate, the next step in processing will either be inlet compression to an "interstage" pressure, typically 300 – 400 psig, or be dew point control and natural gas liquid recovery. Water dew point control is required to meet specifications and to control hydrate formation. Gas hydrate formation is a major concern for engineers in pipeline and natural gas transpor-tation industries as it causes choking/plugging of pipelines and other related problems. Methods of preventing hydrate formation in the plant include lowering the hydrate formation temperature with chemical inhibition or dehydration to re-move the water.

Hydrocarbon dew point or hydrocarbon liquid recovery involves cooling the gas and condensing out the liquids. Hydrocarbon dew point control can be either dehydration followed by cooling/condensation or by a combination of in-hibition/cooling/condensation processes. Refrigeration is performed either by autorefrigeration due to a pressure drop across a valve or by an external mechanical refrigeration process. The temperature to which the gas is cooled depends on whether it is necessary to meet a sales gas hydrocarbon dew point specification or whether substantial liquid recovery is desired. Three situations motivate maximum condensate recovery. The first is the desire to maximize condensate production when processing associated gas. The second situation occurs when processing retrograde condensate gas; here the objective is to recover the condensate and reinject the gas into the formation. Third, in some markets the **natural gas liquids** (NGLs) produced from the condensate may be more valuable as liquid products than as sales gas components, i. e. , their recovery will yield a better profit. Whether to leave maximum NGLs in the gas stream (but still attaining sales hydrocarbon dew point specification) or to recover them as liquids is purely an economic decision made by comparing their value as heat versus the equivalent value as **liquid chemical feedstock**. If the equivalent liquid value is lower than the gas, NGLs should be left in the gas to the extent as possible. However, if the equivalent liquid value is higher than the gas value, then liquid recovery should be maximized.

If gas is produced at lower pressures than typical sales pipeline pressure (approximately 700 –1000 psig), it is compressed to sales gas pressure. Transport of sales gas is done at high pressure in order to reduce pipeline diameter. Pipelines may operate at very high pressures (above 1000 psig) to keep the gas in the dense phase thus preventing condensation and twophase flow. Compression typically requires two to three stages to attain sales gas pressure.

Where there is no available gas pipeline, separated associated gas may be flared. The ability to flare depends on regulations as well as the field location. Increasingly in such cases, separated gas is being conserved by compression and reinjection into producing formations for eventual recovery and sales. Also, in **gas condensate reservoirs**, the gas is often reinjected, or "cycled," to enable higher net recovery of valuable liquid hydrocarbons from the reservoir.

Words and Expressions

condensate	凝析油
noncondensable	非凝析物

impurity	杂质
surge	浪涌
trip	滑脱
dissipate	分散
turbulence	紊流
mercaptan	硫醇

Phrases and Expressions

acid gas	酸性气体
elevational change	高度变化
high velocities and momentum	高速和高动量
slug catcher	段塞捕集器
stratified flow	层流
gravity segregation	重力分离
Reid vapor pressure	雷德蒸汽压
carbonic acid	碳酸
natural gas liquid	天然气凝析液
liquid chemical feedstock	液体化学原料
gas condensate reservoir	凝析气藏

Language Focus

1. Inlet gas receiving is complicated by the fact that transmission lines supplying the plant typically operate with two or three phases present and consequently liquid slugging is common.

(参考译文:进气端的情况很复杂,给处理厂送气的集气管道中流体一般都为两相或三相,因此常有液体段塞流。)

本句中"inlet gas receiving"是主语,"receiving"是动名词;"that"引导同位语从句,指代"the fact",从句中"supplying the plant"是现在分词短语修饰"transmission lines"。

2. Gas pipelines have typically used slug catchers to dissipate the energy of the liquid slugs, to minimize turbulence, to ensure that the gas and liquid flow rates are low enough so that the stratified flow regime and subsequently gravity segregation can occur.

(参考译文:气体管道中使用段塞捕集器来分散液体段塞的能量,减少紊流,降低气体和液体的流速,使流态保持层流,产生重力分离。)

本句中"gas pipelines"为主语,"have typically used"为谓语,"slug catchers"为宾语,后面跟有三个并列的不定式短语做目的状语:"to dissipate the energy of the liquid slugs"、"to minimize turbulence"和"to ensure that the gas and liquid flow rates are low enough so that the stratified flow regime and subsequently gravity segregation can occur"。最后一个不定式中"that"引导宾语从句,其中又包含一个目的状语从句"so that…",意为"以便,所以"。

3. Stabilized liquid will generally have a vapor pressure specification (Reid vapor pressure of < 10 psi), as the product will be injected into a pipeline or transport pressure vessel, which has definite pressure limitations.

(参考译文:稳定处理后的液体要符合蒸汽压指标(雷德蒸汽压小于 10psi),因为要注入有一定压力限制的管道或运输压力罐。)

本句中"as"引导原因状语从句"the product will be injected into a pipeline or transport pressure vessel";"which"引导非限制性定语从句,指代前面的"pipeline or transport pressure vessel"。

Reinforced Learning

I. Answer the following questions for a comprehension of the text.

1. What is the objective of gas processing?

2. What is the first unit module?

3. What acid gas should be treated in natural gas processing?

4. In what situations is maximum condensate recovery motivated?

5. What should be done if gas is produced at lower pressures than typical sales pipeline pressure?

II. Multiple choice: choose the correct one from the alternative answers to give the exact meaning of the words.

1. The first unit module is the physical separation of the distinct phases, which are typically gas, liquid hydrocarbons, liquid water, and/or solids.

A. same　　B. following　　C. different　　D. whole

2. Slugs are normally formed from elevational changes in the inlet supply pipes, changes in gas supply flow rates, and changes in pressure and temperature during transmission.

A. height　　B. elevated　　C. chemical　　D. thermal

3. Slug flow may even be <u>encountered</u> in horizontal pipes under steady – state conditions if the flow regime is not properly selected.

  A. over   B. met   C. worse   D. improved

4. Unstabilized condensates contain a large percentage of methane and ethane, which will <u>vaporize</u> easily in storage tanks.

  A. boil   B. vanish   C. diffuse   D. steam

5. Stabilized liquid will generally have a vapor pressure <u>specification</u> (Reid vapor pressure of < 10 psi) , as the product will be injected into a pipeline or transport pressure vessel, which has definite pressure limitations.

  A. standard B. figure   C. format   D. particular

6. In addition to heavy hydrocarbons and water vapor, natural gas often contains other <u>contaminants</u> that may have to be removed.

  A. matters B. pollutants C. impurities D. chemicals

7. The second situation occurs when processing <u>retrograde</u> condensate gas; here the objective is to recover the condensate and reinject the gas into the formation.

  A. inhibiting B. cooling   C. retroactive D. associated

8. If gas is produced at lower pressures than <u>typical</u> sales pipeline pressure (approximately 700 – 1000 psig) , it is compressed to sales gas pressure.

  A. distinctive B. model   C. general   D. universal

9. Where there is no <u>available</u> gas pipeline, separated associated gas may be flared.

  A. economic B. compressed C. eventual   D. obtainable

10. The ability to flare depends on <u>regulations</u> as well as the field location.

  A. reservations B. conservation C. rules   D. scale

III. Multiple choice: read the four suggested translations and choose the best answer.

1. Each <u>module</u> consists of a single piece or a group of equipment performing a specific function.

  A. 组件   B. 样本   C. 模块   D. 范式

2. Gas pipelines have typically used slug catchers to dissipate the energy of the liquid slugs, to minimize turbulence, to ensure that the gas and liquid flow rates are low enough so that the <u>stratified flow</u> regime and subsequently gravity

segregation can occur.

    A. 层流   B. 平流   C. 分流   D. 流态

    3. The liquids that collect in the <u>slug catcher flow</u> to a three – phase separa-tor from which the two liquid phases, hydrocarbon condensate and water/metha-nol or water/glycol phases, are outputs.

    A. 堵塞疏通器流体   B. 堵塞疏通后的流出液体

    C. 段塞捕集器分离出的液体   D. 段塞捕集器疏导的液体

    4. Hydrocarbon condensate recovered from natural gas may be shipped without further processing but is typically stabilized to produce a <u>safe transporta-ble liquid</u>.

    A. 安全便携式液体   B. 能够安全运输的液体

    C. 既安全又轻便的液体   D. 安全可靠的凝析液

    5. Stabilized liquid will generally have a vapor pressure specification (Reid vapor pressure of < 10 psi), as the product will be injected into a pipeline or transport pressure vessel, which has <u>definite pressure limitations</u>.

    A. 确切的压力限制   B. 严格符合要求的蒸汽压指标

    C. 一定的压强限度   D. 坚决达到气压限制标准

    6. Carbon dioxide (CO_2), hydrogen sulfide (H_2S), and other sulfur – con-taining species such as <u>mercaptans</u> are compounds that require complete or par-tial removal.

    A. 硫化氢   B. 碳酸   C. 硫醇   D. 酸性气体

    7. Natural gas with H_2S or other sulfur compounds present is called "sour gas," whereas gas with only CO_2 is called "<u>sweet</u>."

    A. 弱硫酸 碳酸   B. 酸气 甜气

    C. 酸性气体 甜性气体   D. 酸味气体 甜味气体

    8. Depending on the pressure at the plant gate, the next step in processing will either be inlet compression to an "<u>interstage</u>" pressure, typically 300 – 400 psig, or be dew point control and natural gas liquid recovery.

    A. 中间压力   B. 门限压力   C. 中层压力   D. 中级压力

    9. Hydrocarbon dew point control can be either dehydration followed by cooling/condensation or by a combination of <u>inhibition/cooling/condensation</u> processes.

    A. 抑制/冷却/凝析   B. 脱水/冷却/凝析

    C. 回收/冷却/压缩   D. 控制/降温/冷凝

10. Increasingly in such cases, separated gas is being conserved by compression and reinjection into <u>producing formations</u> for eventual recovery and sales.

    A. 生产形式   B. 生产结构   C. 产气队   D. 产气层

Ⅳ. Put the following sentences into Chinese.

1. The first unit module is the physical separation of the distinct phases, which are typically gas, liquid hydrocarbons, liquid water, and/or solids.

2. The arrival of "slugs" at production or processing equipment impacts the operation of production facilities negatively, causing both mechanical problems (due to high velocities and momentum) and process problems (increasing liquid levels, causing surges and trips).

3. Unstabilized condensates contain a large percentage of methane and ethane, which will vaporize easily in storage tanks.

4. Gas hydrate formation is a major concern for engineers in pipeline and natural gas transportation industries as it causes choking/plugging of pipelines and other related problems.

5. Pipelines may operate at very high pressures to keep the gas in the dense phase thus preventing condensation and twophase flow.

Ⅴ. Put the following paragraphs into Chinese.

1. Raw natural gas after transmission through the field – gathering network must be processed before it can be moved into long – distance pipeline systems for use by consumers. The objective of gas processing is to separate natural gas, condensate, noncondensable, acid gases, and water from a gas – producing well and condition these fluids for sale or disposal. In some cases, little processing is needed; however, most natural gas requires processing equipment at the gas processing plant to remove impurities, water, and excess hydrocarbon liquid and to control delivery pressure.

2. Hydrocarbon dew point or hydrocarbon liquid recovery involves cooling the gas and condensing out the liquids. Hydrocarbon dew point control can be either dehydration followed by cooling/condensation or by a combination of inhibition/cooling/condensation processes. Refrigeration is performed either by autorefrigeration due to a pressure drop across a valve or by an external mechanical refrigeration process. The temperature to which the gas is cooled de-

pends on whether it is necessary to meet a sales gas hydrocarbon dew point specification or whether substantial liquid recovery is desired. Whether to leave maximum NGLs in the gas stream (but still attaining sales hydrocarbon dew point specification) or to recover them as liquids is purely an economic decision made by comparing their value as heat versus the equivalent value as liquid chemical feedstock. If the equivalent liquid value is lower than the gas, NGLs should be left in the gas to the extent as possible. However, if the equivalent liquid value is higher than the gas value, then liquid recovery should be maximized.

4.2   Gravity Separation

📖 Guidance to Reading

*In physical separation of gas and liquids or solids, three principles often used are **momentum**, **gravity settling**, and **coalescing**, among which gravity segregation is the main force that accomplishes the separation. Gravity separators are pressure vessels that separate a mixed – phase stream into gas and liquid phases that are relatively free of each other. Gravity separators are often classified as vertical and horizontal by their **geometrical configuration**; and two – phase/three – phase separators by their function. Separators are sometimes called "**scrubbers**" when the ratio of gas rate to liquid rate is very high.*

📖 Text

Three principles used to achieve physical separation of gas and liquids or solids are momentum, gravity settling, and coalescing. Any separator may employ one or more of these principles; however, the fluid phases must be **immiscible** and have different densities for separation to occur. Momentum force is utilized by changing the direction of flow and is usually employed for bulk separation of the fluid phases. The gravitational force is utilized by reducing velocity so the liquid droplets can settle out in the space provided. Gravity segregation is the main force that accomplishes the separation, which means the heaviest fluid settles to the bottom and the lightest fluid rises to the top. However, very small droplets such as mist cannot be separated practically by gravity. These droplets can be **coalesced** to form larger droplets that will settle by gravity.

Gravity Separators

Gravity separators are pressure vessels that separate a mixed – phase stream into gas and liquid phases that are relatively free of each other. In a gravity separator, gravitational forces control separation, and the efficiency of the gas/liquid separation is increased by lowering the gas velocity. Gravity separators are often classified by their geometrical configuration (vertical, horizontal) and by their function (two – phase/three – phase separator). Separators are sometimes called "scrubbers" when the ratio of gas rate to liquid rate is very high. These vessels usually have a small liquid collection section and are recommended only for the following items.

(1) Secondary separation to remove carryover fluids from process equipment such as **absorbers** and **liquid dust scrubbers**.

(2) Gas line separation downstream from a separator and where flow lines are not long.

(3) **Miscellaneous** separation where the gas – liquid ratio is extremely high.

In any case, these equipments have the same configuration and are sized in accordance with the same procedure of separators.

Components and Features

All gravity separators normally have the following components or features.

(1) A primary gas/liquid separation section with an **inlet divertor** to remove the bulk of the liquid from the gas.

(2) A gravity – settling section providing adequate retention time so that proper settling may take place.

(3) A **mist extractor** at the gas outlet to capture **entrained droplets** or those too small to settle by gravity.

(4) Proper pressure and liquid – level controls.

Gravity separators are designed as either horizontal or vertical pressure vessels. Figure 4. 2. 1 is a typical scheme of a three – phase horizontal separator.

The fluid enters the separator and hits an inlet diverter. This sudden change in momentum generates the initial bulk separation of liquid and gas. In most designs, the inlet diverter contains a downcomer that directs the liquid flow below the oil/water interface. This forces the inlet mixture of oil and water to mix with the water continuous phase in the bottom of the vessel and rise through the oil/

water interface. This process is called "water washing" and promotes the coalescence of water droplets that are entrained in the oil continuous phase. The inlet diverter assures that little gas is carried with the liquid, and the water wash assures that the liquid does not fall on top of the gas/oil or oil/water interface, mixing the liquid retained in the vessel and making control of the oil/water interface difficult.

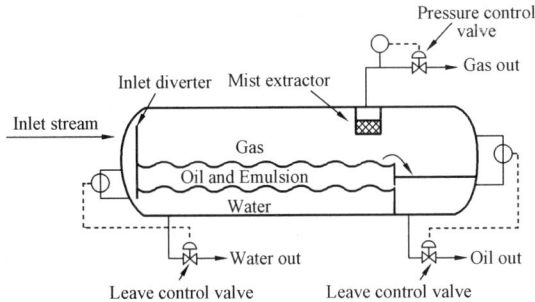

Figure 4.2.1　A typical scheme of a horizontal three - phase separator

The liquid - collecting section of the vessel provides sufficient time so that the oil and emulsion form a layer or "oil pad" at the top. The free water settles to the bottom. The produced water flows from a **nozzle** in the vessel located upstream of the **oil weir**. An interface level controller senses the height of the oil/water interface. The controller sends a signal to the water dump valve, thus allowing the correct amount of water to leave the vessel so that the oil/water interface is maintained at the design height. The gas flows horizontally and outs through a mist extractor to a pressure control valve that maintains constant vessel pressure.

Figure 4.2.2 shows a typical configuration for a vertical three - phase separator. In the vertical separator, the flow enters the vessel through the side as in the horizontal separator and the inlet diverter separates the bulk of the gas. The gas moves upward, usually passing through a mist extractor to remove suspended mist, and then the dry gas flows out. A downcomer is required to transmit the liquid collected through the oil - gas interface so as not to disturb the oil - skimming action taking place.

A chimney is needed to equalize gas pressure between the lower section and the gas section. The spreader or downcomer outlet is located at the oil - water interface. From this point as the oil rises any free water trapped within the

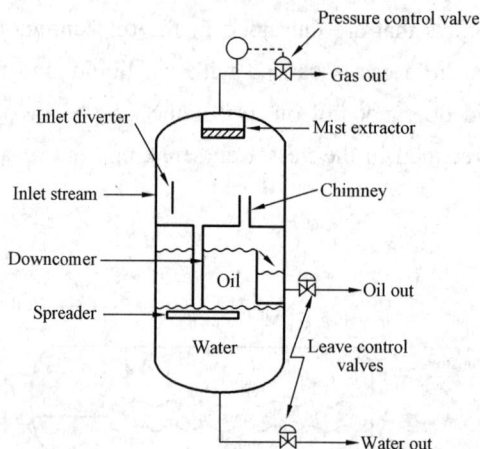

Figure 4.2.2   A typical scheme of a vertical three – phase separator

oil phase separates out. The water droplets flow countercurrent to the oil. Similarly, the water flows downward and oil droplets trapped in the water phase tend to rise countercurrent to the water flow.

It should be clear that the principles of operation of three – phase vertical separators are the same as the three – phase horizontal separators described earlier. Essentially, the only difference is that horizontal separators have separation acting **tangentially** to flow, whereas vertical separators have separation acting parallel to flow. In the vertical separator, level control is not also critical, where the liquid level can **fluctuate** several inches without affecting operating efficiency. However, it can affect the pressure drop for the downcomer pipe (from the **demister**), therefore affecting demisting device drainage.

Separator Selection

There are no simple rules for separator selection. Sometimes, both configurations should be evaluated to decide which is more economical.

Horizontal Separators

Horizontal separators are used most commonly in the following conditions.

(1) Large volumes of gas and/or liquids.

(2) **High – to – medium gas/oil ratio (GOR)** streams.

(3) **Foaming crudes**.

(4) Three – phase separation.

Advantages

(1) Require smaller diameter for similar gas capacity as compared to vertical vessels.

(2) No counterflow (gas flow does not oppose drainage of mist extractor).

(3) Large liquid surface area for foam dispersion generally reduces turbulence.

(4) Larger surge volume capacity.

Disadvantages

(1) Only part of shell available for passage of gas.

(2) Occupies more space unless "stack" mounted.

(3) Liquid level control is more critical.

(4) More difficult to clean produced sand, mud, wax, paraffin, etc.

Vertical Separators

These separators are used in the following conditions.

(1) Small flow rates of gas and/or liquids.

(2) Very high GOR streams or when the total gas volumes are low.

(3) Plot space is limited.

(1) Ease of level control is desired.

Advantages

(1) Liquid level control is not so critical.

(2) Have good bottom – drain and clean – out facilities.

(3) Can handle more sand, mud, paraffin, and wax without plugging.

(4) Less tendency for reentrainment.

(5) Has full diameter for gas flow at top and oil flow at bottom.

(6) Occupies smaller plot area.

Disadvantages

(1) Require larger diameter for a given gas capacity, therefore, most competitive for very low GOR or very high GOR or **scrubber applications**.

(2) Not recommended when there is a large slug potential.

(3) More difficult to reach and service top – mounted instruments and safety devices.

⌐⊏ **Words and Expressions**

momentum	动量学
scrubber	净气器
coalescing	聚结
immiscible	不相溶的
coalesce	聚结
absorber	吸收器
miscellaneous	混杂的
nozzle	喷嘴
tangentially	切线地
fluctuate	波动
demister	除雾器

⌐⊏ **Phrases and Expressions**

gravity settling	重力沉降
geometrical configuration	几何结构
liquid dust scrubber	液体除尘器
inlet divertor	入口分流器
mist extractor	除雾器
entrained droplet	夹带液滴
oil weir	挡油板
high – to – medium gas/oil ratio GOR	较高—中等气油比
foaming crude	泡沫油
scrubber application	设备清洗

⌐⊏ **Language Focus**

1. Gravity segregation is the main force that accomplishes the separation, which means the heaviest fluid settles to the bottom and the lightest fluid rises to the top.

（参考译文：重力是完成分离最主要的力，最重的液体沉降到底部，而最轻的升到顶部。）

本句中"that"引导限制性定语从句"accomplishes the separation"，指代"the main force"；"which"引导非限制性定语从句"means the heaviest fluid settles to the bottom and the lightest fluid rises to the top",指代前面整个句子。

2. The inlet diverter assures that little gas is carried with the liquid, and the water wash assures that the liquid does not fall on top of the gas/oil or oil/water interface, mixing the liquid retained in the vessel and making control of the oil/water interface difficult.

（参考译文：入口分流器确保液体中基本不含气体，并且水洗确保液体不落在油气或油水界面上部，否则与容器内液体混合后使油水界面很难控制。）

本句中"and"连接两个并列的分句"The inlet diverter assures that little gas is carried with the liquid"和"the water wash assures that the liquid does not fall on top of the gas/oil or oil/water interface, mixing the liquid retained in the vessel and making control of the oil/water interface difficult";两个"that"分别引导宾语从句"little gas is carried with the liquid"和"the liquid does not fall on top of the gas/oil or oil/water interface";"mixing the liquid retained in the vessel and making control of the oil/water interface difficult"为伴随状语。

3. Essentially, the only difference is that horizontal separators have separation acting tangentially to flow, whereas vertical separators have separation acting parallel to flow.

（参考译文：本质上唯一不同的是，水平分离器是切线流，而垂直分离器是平行流。）

本句中"whereas"作连词，意为"然而"，表示"horizontal separators"与"vertical separators"的不同点。

⌗ Reinforced Learning

Ⅰ. Answer the following questions for a comprehension of the text.

1. What are the principles used to achieve physical separation of gas and liquids or solids?

2. What is the recommended use of gravity separators?

3. What are the features of gravity separators?

4. What is the difference between three – phase vertical separators and three – phase horizontal separators essentially?

5. In what conditions are horizontal separators used most commonly?

II. Multiple choice: choose the correct one from the alternative answers to give the exact meaning of the words.

1. Three <u>principles</u> used to achieve physical separation of gas and liquids or solids are momentum, gravity settling, and coalescing.

    A. laws   B. principals   C. separators   D. characters

2. Gravity separators are pressure vessels that separate a mixed – phase stream into gas and liquid phases that are <u>relatively</u> free of each other.

    A. somewhat   B. related   C. generally   D. particularly

3. Separators are sometimes called "scrubbers" when the <u>ratio</u> of gas rate to liquid rate is very high.

    A. versus   B. comparison   C. proportion   D. percentage

4. Figure 4. 2. 1 is a typical <u>scheme</u> of a three – phase horizontal separator.

    A. picture   B. figure   C. symbol   D. plan

5. A chimney is needed to <u>equalize</u> gas pressure between the lower section and the gas section.

    A. distribute   B. arrange   C. balance   D. compress

6. In the vertical separator, level control is not also <u>critical</u>, where the liquid level can fluctuate several inches without affecting operating efficiency.

    A. necessary   B. crucial   C. optimal   D. indispensable

7. However, it can affect the pressure drop for the downcomer pipe (from the demister), therefore affecting <u>demisting</u> device drainage.

    A. dismisting   B. mist distilling

    C. mist removing   D. mist vaporation

8. Sometimes, both configurations should be <u>evaluated</u> to decide which is more economical.

    A. decided   B. checked   C. accessed   D. assessed

9. Sometimes, both configurations should be evaluated to decide which is <u>more economical</u>.

    A. more economic   B. better

    C. more enduring   D. more cost – effective

10. Require larger diameter for a given gas capacity, therefore, most competitive for very low GOR or very high GOR or <u>scrubber applications</u>.

    A. equipment cleaning   B. washing machinery

    C. washing machine   D. facility clearing

Ⅲ. Multiple choice：read the four suggested translations and choose the best answer.

1. Gravity separators are often classified by their geometrical configuration (vertical,horizontal) and by their function (two – phase/three – phase separator).

　　A. 数学结构　　B. 结构模型　　C. 几何构造　　D. 几何结构

2. The liquid – collecting section of the vessel provides sufficient time so that the oil and emulsion form a layer or "oil pad" at the top.

　　A. 液体集合区　　　　　　　B. 采集液体部
　　C. 集液区　　　　　　　　　D. 集液部

3. The produced water flows from a nozzle in the vessel located upstream of the oil weir.

　　A. 喷鼻　　　B. 喷嘴　　　　C. 排口　　　　　D. 扇口

4. The gas moves upward,usually passing through a mist extractor to remove suspended mist,and then the dry gas flows out.

　　A. 悬浮雾流　　B. 中断薄雾　　　C. 阻断雾气　　　D. 去除雾气

5. A downcomer is required to transmit the liquid collected through the oil – gas interface so as not to disturb the oil – skimming action taking place.

　　A. 略油　　　B. 去油　　　　C. 排油　　　　　D. 撇油

6. However,it can affect the pressure drop for the downcomer pipe (from the demister),therefore affecting demisting device drainage.

　　A. 防雾器　　B. 除霜器　　　C. 除雾器　　　　D. 降雾器

7. Large liquid surface area for foam dispersion generally reduces turbulence.

　　A. 泡沫扩散　　B. 水泡分散　　　C. 白泡分散　　　D. 泡沫散开

8. No counterflow (gas flow does not oppose drainage of mist extractor).

　　A. 排除　　　B. 排放　　　　C. 排驱　　　　　D. 排污

9. More difficult to clean produced sand,mud,wax,paraffin,etc.

　　A. 砂、泥、重烃、水蒸气　　　B. 砂、泥浆、蜡、石蜡
　　C. 沙、气体、液态烃、液态水　D. 沙、硫化氢、甲醇、硫醇

10. More difficult to reach and service top – mounted instruments and safety devices.

　　A. 安装在顶层的工具　　　　B. 顶部仪器
　　C. 顶上的器具　　　　　　　D. 顶部连接的器具

— 133 —

IV. Put the following sentences into Chinese.

1. In a gravity separator, gravitational forces control separation, and the efficiency of the gas/liquid separation is increased by lowering the gas velocity.

2. The controller sends a signal to the water dump valve, thus allowing the correct amount of water to leave the vessel so that the oil/water interface is maintained at the design height.

3. In the vertical separator, the flow enters the vessel through the side as in the horizontal separator and the inlet diverter separates the bulk of the gas.

4. It should be clear that the principles of operation of three – phase vertical separators are the same as the three – phase horizontal separators described earlier.

5. In the vertical separator, level control is not also critical, where the liquid level can fluctuate several inches without affecting operating efficiency.

V. Put the following paragraphs into Chinese.

1. Three principles used to achieve physical separation of gas and liquids or solids are momentum, gravity settling, and coalescing. Any separator may employ one or more of these principles; however, the fluid phases must be immiscible and have different densities for separation to occur. Momentum force is utilized by changing the direction of flow and is usually employed for bulk separation of the fluid phases. The gravitational force is utilized by reducing velocity so the liquid droplets can settle out in the space provided. Gravity segregation is the main force that accomplishes the separation, which means the heaviest fluid settles to the bottom and the lightest fluid rises to the top. However, very small droplets such as mist cannot be separated practically by gravity. These droplets can be coalesced to form larger droplets that will settle by gravity.

2. A chimney is needed to equalize gas pressure between the lower section and the gas section. The spreader or downcomer outlet is located at the oil – water interface. From this point as the oil rises any free water trapped within the oil phase separates out. The water droplets flow countercurrent to the oil. Similarly, the water flows downward and oil droplets trapped in the water phase tend to rise countercurrent to the water flow.

4.3　Multistage Separation

Guidance to Reading

*Multistage separation is applied to achieve good separation between gas and liquid phases and maximizing hydrocarbon liquid recovery. The number of stages normally ranges between two and four, depending on the **gas/oil ratio** and the well stream pressure. The main objective of stage separation is to provide maximum stabilization to the **resultant phases** (gas and liquid) leaving the final separator, which means that the considerable amounts of gas or liquid will not evolve from the final liquid and gas phases, respectively.*

Text

Multistage Separation

To achieve good separation between gas and liquid phases and maximizing hydrocarbon liquid recovery, it is necessary to use several separation stages at decreasing pressures in which the well stream is passed through two or more separators arranged in series. The operating pressures are **sequentially** reduced, hence the highest pressure is found at the first separator and the lowest pressure at the final separator. In practice, the number of stages normally ranges between two and four, which depends on the gas/oil ratio (GOR) and the well stream pressure, where two – stage separation is usually used for low GOR and low well stream pressure, three – stage separation is used for medium to high GOR and intermediate inlet pressure, and four – stage separation is used for high GOR and a high pressure well stream. Note that three – stage separation usually represents the economic optimum, where it allows 2% to 12% higher liquid recovery in comparison with two – stage separation and, in some cases, recoveries up to 25% higher. To recover the gases fractions produced in the separators operating at medium pressure and low pressure, it is necessary to recompress them to the pressure of the high – pressure separator. However, for an associated gas, recompression is sometimes considered too costly; hence the gas produced from the low – pressure separator may be flared.

It should be noted that the main objective of stage separation is to provide maximum stabilization to the resultant phases (gas and liquid) leaving the final

separator, which means that the considerable amounts of gas or liquid will not e-
volve from the final liquid and gas phases, respectively.

Centrifugal Separators

In centrifugal or **cyclone separators**, **centrifugal forces** act on droplet at
forces several times greater than gravity as it enters a **cylindrical separator**. This centrifugal force can range from 5 times gravity in large, low – velocity units to 2000 times gravity in small, high – pressure units. Generally, centrifugal separators are used for removing droplets greater than $100\mu m$ in diameter,
and a properly sized centrifugal separator can have a reasonable **removal efficiency** of droplet sizes as low as $10\mu m$.

Centrifugal separators are also extremely useful for gas streams with **high
particulate loading**. The compact dimensions, smaller footprint, and lower
weight of the GLCC have a potential for cost savings to the industry, especially
in offshore applications. Also, the GLCC reduces the inventory of hydrocarbons
significantly, which is critical to environmental and safety considerations. The
GLCC separator, used mainly for bulk gas/liquid separation, can be designed for
various levels of expected performance.

Twister Supersonic Separator

The Twister supersonic separator is a unique combination of known physical processes, combining expansion, **cyclonic gas**/liquid separation, and recompression process steps in a compact, **tubular device** to condense and separate
water and heavy hydrocarbons from natural gas. Condensation and separation at
supersonic velocity are key to achieving step – change reductions in both capital
and operating costs. The residence time inside the Twister supersonic separator
is only milliseconds, allowing hydrates no time to form and avoiding the requirement for hydrate inhibition chemicals. Elimination of the associated chemical regeneration systems avoids harmful **benzene, toluene**, and **xylene** emissions
to the environment or the expense of chemical recovery systems. The simplicity
and reliability of a static device, with no rotating parts, which operates without
chemicals, ensure a simple facility with a high availability suitable for **unmanned operation** in harsh and/or offshore environments. In addition, the compact and low weight Twister system design enables **debottlenecking** of existing
space and weight – constrained platforms.

Slug Catchers

Slug catchers are used at the **terminus** of offshore pipelines to catch large slugs of liquid in pipelines, to hold these slugs temporarily, and then to allow them to follow into downstream equipment and facilities at a rate at which the liquid can be handled properly. Slug catchers may be either a vessel or constructed pipes. Pipe – type slug catchers are frequently less expensive than vessel type slug catchers of the same capacity due to thinner wall requirements of smaller pipe diameter. The manifold nature of multiple pipe – type slug catchers also makes possible the later addition of additional capacity by laying more parallel pipes. The general configuration of a pipe – type slug catcher consists of the following parts.

(1) Fingers with dual slope and three distinct sections: gas/liquid separation, intermediate, and storage sections.

(2) Gas risers connected to each finger at the transition zone between separation and intermediate sections.

(3) Gas equalization lines located on each finger. These lines are located within the slug storage section.

(4) Liquid header collecting liquid from each finger. This header will not be sloped and is configured **perpendicular** to the fingers.

Note that it has been assumed that all liquids (condensate and water) are collected and sent to an inlet three – phase separator, although it is possible to separate condensate and water at the fingers directly. When doing condensate/water separation at the slug catcher itself, we have to allow separately for the maximum condensate slug and the maximum water slug in order to ensure continuous level control.

Separation of gas and liquid phases is achieved in the first section of the fingers. The length of this section promotes a stratified flow pattern and permits primary separation to occur. Ideally, liquid droplets, 600 μm and below, are removed from the gas disengaged into the gas risers, which are located at the end of this section. The length of the intermediate section is minimal such that there is no liquid level beneath the gas riser when the slug catcher is full, i. e. , storage section completely full. This section comprises a change in elevation between the gas risers and the storage section that allows a clear distinction between liquid and gas phases.

The length of the storage section ensures that the maximum slug volume can be retained without liquid carryover in the gas outlet. During normal operations, the normal liquid level is kept at around the top of the riser from each finger into the main liquid collection header, which is equivalent to approximately a 5 – min operation of the condensate stabilization units at maximum capacity.

Slug catcher design is dependent on several factors, of which the most important are pigging operation and changes in flow rates. Pigging can also be used as a means of limiting the required slug catcher size, where by pigging at frequent intervals, liquid inventory buildup in a pipeline can be reduced and the maximum slug size can be limited. However, a slug catcher size should be chosen on the basis of balancing the cost of frequent pigging operations and the capital reduction of smaller slug catchers.

Words and Expressions

sequentially	循序地
benzene	苯
toluene	甲苯
xylene	二甲苯
debottlenecking	消除瓶颈
terminus	末端
perpendicular	垂直的

Phrases and Expressions

gas/oil ratio	气油比
resultant phase	最终相
centrifugal separator	离心分离器
cyclone separator	旋风分离器
centrifugal force	离心力
cylindrical separator	圆柱形分离器
removal efficiency	分离效率
high particulate loading	携带颗粒较多
twister supersonic separator	扭转式超声分离器
cyclonic gas	旋流式气
tubular device	管状装置

unmanned operation　　　　　　　　无人操作

slug catcher　　　　　　　　　　　段塞流捕集器

Language Focus

1. To achieve good separation between gas and liquid phases and maximizing hydrocarbon liquid recovery, it is necessary to use several separation stages at decreasing pressures in which the well stream is passed through two or more separators arranged in series.

（参考译文：为了使气液两相更好地分离，以便使烃类液体回收率最大化，要采取多级降压分离，即让井流物通过两个或两个以上的系列分离器。）

本句中"To achieve good separation between gas and liquid phases and maximizing hydrocarbon liquid recovery"是目的状语；"it"是形式主语，真正的主语是"to use several separation stages at decreasing pressures"；"in which"引导限制性定语从句，指代"in several separation stages"。

2. The simplicity and reliability of a static device, with no rotating parts, which operates without chemicals, ensure a simple facility with a high availability suitable for unmanned operation in harsh and/or offshore environments.

（参考译文：这种简单可靠的固定装置无旋转部件，不用加入化学物质，非常适合在恶劣环境或近海环境中广泛进行无人操作。）

本句主谓宾结构是"the simplicity and reliability of a static device ensure a simple facility"，"with no rotating parts, which operates without chemicals"为插入语，补充说明主语，"which"引导非限制性定语从句；"suitable for unmanned operation in harsh and/or offshore environments"是形容词短语作后置定语修饰"a simple facility with a high availability"。

3. Slug catchers are used at the terminus of offshore pipelines to catch large slugs of liquid in pipelines, to hold these slugs temporarily, and then to allow them to follow into downstream equipment and facilities at a rate at which the liquid can be handled properly.

（参考译文：段塞流捕集器用于近海管道末端来捕获管道中的液体大段塞，并暂时储存这些段塞，然后使其以适当流速进入下游设备和处理设施。）

本句中有三个不定式短语"to catch large slugs of liquid in pipelines"、"to hold these slugs temporarily"和"to allow them to follow into downstream equipment and facilities at a rate at which the liquid can be handled properly"为并列目的状语;"at which"引导限制性定语从句,相当于"at a rate"。

🔲 Reinforced Learning

I. Answer the following questions for a comprehension of the text.

1. What is the main objective of stage separation?

2. What is the use of centrifugal separators for gas streams with high particulate loading?

3. What is the general configuration of a pipe – type slug catcher?

4. How does a twister supersonic separator work?

5. What are the most important factors of slug catcher design?

II. Multiple choice:choose the correct one from the alternative answers to give the exact meaning of the words.

1. In practice,the number of stages normally ranges between two and four, which depends on the gas/oil ratio (GOR) and the well stream pressure,where two – stage separation is usually used for low GOR and low well stream pressure,three – stage separation is used for medium to high GOR and intermediate inlet pressure,and four – stage separation is used for high GOR and a high pressure well stream.

A. high   B. middle   C. low   D. mediating

2. The GLCC separator,used mainly for bulk gas/liquid separation,can be designed for various levels of expected performance.

A. little   B. trunk   C. mass   D. bunch

3. The residence time inside the Twister supersonic separator is only milliseconds,allowing hydrates no time to form and avoiding the requirement for hydrate inhibition chemicals.

A. request   B. inquiry   C. must   D. need

4. Elimination of the associated chemical regeneration systems avoids harmful benzene,toluene, and xylene emissions to the environment or the expense of chemical recovery systems.

A. riddance   B. limitation   C. compression D. decrease

5. Pipe – type slug catchers are frequently less expensive than vessel type slug catchers of the same capacity due to thinner wall requirements of smaller

pipe diameter.

A. more or less　B. repeatedly　　C. seldom　　　D. not so often

6. The manifold nature of multiple pipe – type slug catchers also makes possible the later addition of additional capacity by laying more parallel pipes.

A. many　　　　B. manmade　　C. unfolding　　D. intermediate

7. Ideally, liquid droplets, 600 μm and below, are removed from the gas disengaged into the gas risers, which are located at the end of this section.

A. Perfectly　　B. Generally　　C. Hopefully　　D. Fortunately

8. The length of the intermediate section is minimal such that there is no liquid level beneath the gas riser when the slug catcher is full, i. e. , storage section completely full.

A. miniature　　B. maximum　　　C. intermediate　D. minimum

9. The length of the storage section ensures that the maximum slug volume can be retained without liquid carryover in the gas outlet.

A. carriage　　　B. carrion　　　　C. conveyance　D. carrier

10. Pigging requirements and ramp – up periods are determined by transient analysis once the steady – state liquid holdup of the pipeline is understood.

A. resistant　　B. instant　　　　C. transparent　D. constant

Ⅲ. Multiple choice：read the four suggested translations and choose the best answer.

1. It should be noted that the main objective of stage separation is to provide maximum stabilization to the resultant phases (gas and liquid) leaving the final separator, which means that the considerable amounts of gas or liquid will not evolve from the final liquid and gas phases, respectively.

A. 最后阶段　　B. 合成相　　　C. 最终相　　　D. 结果相

2. In centrifugal or cyclone separators, centrifugal forces act on droplet at forces several times greater than gravity as it enters a cylindrical separator.

A. 中心气旋除尘器　　　　　B. 远中心旋风除尘器
C. 向心或旋风分离器　　　　D. 离心或旋风分离器

3. Generally, centrifugal separators are used for removing droplets greater than 100 μm in diameter, and a properly sized centrifugal separator can have a reasonable removal efficiency of droplet sizes as low as 10μm.

A. 分离效率　　B. 除去效率　　C. 消除功效　　D. 去除效能

4. Centrifugal separators are also extremely useful for gas streams with high

particulate loading.

　　A. 高速粒子加载　　　　　　　B. 高度微粒载荷

　　C. 携带高速粒子　　　　　　　D. 携带颗粒较多

　　5. The compact dimensions, smaller footprint, and lower weight of the GL-CC have a potential for cost savings to the industry, especially in offshore applications.

　　A. 尺寸紧凑　　B. 规模压缩　　C. 压缩大小　　D. 缩小规格

　　6. Also, the GLCC reduces the inventory of hydrocarbons significantly, which is critical to environmental and safety considerations.

　　A. 存货　　　　B. 总量　　　　C. 货存　　　　D. 目录

　　7. The Twister supersonic separator is a unique combination of known physical processes, combining expansion, cyclonic gas/liquid separation, and re-compression process steps in a compact, tubular device to condense and separate water and heavy hydrocarbons from natural gas.

　　A. 气旋气　　　B. 飓风气　　　　C. 旋风气　　　D. 旋流式气

　　8. In addition, the compact and low weight Twister system design enables debottlenecking of existing space and weight – constrained platforms.

　　A. 摘下瓶盖　　B. 降解瓶子　　　C. 消除瓶颈　　D. 打碎瓷瓶

　　9. Note that it has been assumed that all liquids (condensate and water) are collected and sent to an inlet three – phase separator, although it is possible to separate condensate and water at the fingers directly.

　　A. 手指　　　　B. 指头　　　　C. 指形管　　　D. 剔除器

　　10. Pigging requirements and ramp – up periods are determined by transient analysis once the steady – state liquid holdup of the pipeline is understood.

　　A. 上升周期　　B. 等待时间　　　C. 扩展时期　　D. 流量激动周期

Ⅳ. Put the following sentences into Chinese.

　　1. Generally, centrifugal separators are used for removing droplets greater than $100\mu m$ in diameter, and a properly sized centrifugal separator can have a reasonable removal efficiency of droplet sizes as low as $10\mu m$.

　　2. The compact dimensions, smaller footprint, and lower weight of the GL-CC have a potential for cost savings to the industry, especially in offshore applications.

　　3. When doing condensate/water separation at the slug catcher itself, we have toallow separately for the maximum condensate slug and the maximum wa-

ter slug in order to ensure continuous level control.

4. During normal operations, the normal liquid level is kept at around the top of the riser from each finger into the main liquid collection header, which is equivalent to approximately a 5 – min operation of the condensate stabilization units at maximum capacity.

5. Pigging can also be used as a means of limiting the required slug catcher size, where by pigging at frequent intervals, liquid inventory buildup in a pipeline can be reduced and the maximum slug size can be limited.

V. Put the following paragraphs into Chinese.

1. To achieve good separation between gas and liquid phases and maximizing hydrocarbon liquid recovery, it is necessary to use several separation stages at decreasing pressures in which the well stream is passed through two or more separators arranged in series. The operating pressures are sequentially reduced, hence the highest pressure is found at the first separator and the lowest pressure at the final separator. In practice, the number of stages normally ranges between two and four, which depends on the gas/oil ratio (GOR) and the well stream pressure, where two – stage separation is usually used for low GOR and low well stream pressure, three – stage separation is used for medium to high GOR and intermediate inlet pressure, and four – stage separation is used for high GOR and a high pressure well stream.

2. The residence time inside the Twister supersonic separator is only milliseconds, allowing hydrates no time to form and avoiding the requirement for hydrate inhibition chemicals. Elimination of the associated chemical regeneration systems avoids harmful benzene, toluene, and xylene emissions to the environment or the expense of chemical recovery systems. The simplicity and reliability of a static device, with no rotating parts, which operates without chemicals, ensure a simple facility with a high availability suitable for unmanned operation in harsh and/or offshore environments.

4.4 Natural Gas Liquids Recovery

Guidance to Reading

Most natural gas is processed to remove the heavier hydrocarbon liquids (NGLs) from the natural gas stream. Recovery of NGL components in gas not only may be required for hydrocarbon dew point control in a natural gas stream, but also yields a source of revenue, as NGLs normally have significantly greater value as separate marketable products than as part of the natural gas stream. Therefore it makes economic sense to recover them separately for maximum gas field monetization. However, regardless of more monetary value, natural gas usually must be processed to meet the specification for safe delivery and combustion.

🔲 Text

The heavier hydrocarbon liquids from natural gas, commonly referred to as **natural gas liquids** (**NGLs**), include ethane, propane, butanes, and **natural gasoline** (**condensate**). Lighter NGL fractions can be sold as fuel or feedstock to refineries and petrochemical plants, while the heavier portion can be used as **gasoline – blending stock**. The price difference between selling NGL as a liquid and as fuel, commonly referred to as the "**shrinkage value**," often dictates the recovery level desired by the gas processors. Regardless of the **economic incentive**, however, gas usually must be processed to meet the specification for safe delivery and combustion. Hence, NGL recovery **profitability** is not the only factor in determining the degree of NGL extraction. The removal of natural gas liquids usually takes place in a relatively centralized processing plant, where the recovered NGLs are then treated to meet commercial specifications before moving into the NGL transportation infrastructure.

NGL Recovery Processes

Figure 4.4.1 shows the phase behavior of a natural gas as a function of pressure and temperature. Obviously any cooling outside the **retrograde condensation** zone will induce condensation and yield NGL. Retrograde condensation phenomenon has an important application in NGL production.

Some plants operate at inlet pressures above the critical point and thus revaporize NGLs when the temperature drops below the retrograde temperature. It is therefore important to know where we are on the phase envelope.

Refrigeration Processes

Refrigeration processes are used at many different temperature levels to condense or cool gases, vapor, or liquids. Mechanical refrigeration is the simplest and most direct process for NGL recovery.

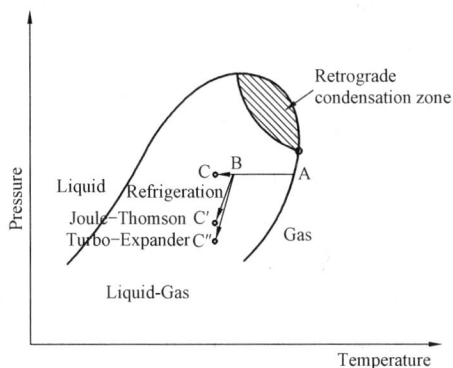

Figure 4. 4. 1　Thermodynamic pathways of different NGL recovery technologies

Lean Oil Absorption

The absorption method of NGL recovery is very similar to using absorption for gas dehydration. The main difference is that, in NGL absorption, an absorbing oil is used as opposed to glycol. This absorbing oil has an "affinity" for NGLs in much the same manner as glycol has an affinity for water.

Lean oil absorption is the oldest and least efficient process to recover NGLs. Note that an oil absorption plant cannot recover ethane and propane effectively, where it requires circulating large amounts of absorption oil, demands attendant maintenance, and consumes too much fuel. Lean oil absorption plants are not as popular as they once were and are rarely, if ever, constructed anymore. They are expensive and more complex to operate, and it is difficult to predict their efficiency at removing liquids from the gas as the lean oil deteriorates with time.

Solid Bed Adsorption

This method uses **adsorbents** that have the capability to adsorb heavy hydrocarbons from natural gas. The adsorbent may be **silica gel** or **activated charcoal**. **Activated alumina** cannot be used in the presence of heavy hydrocarbons, which **foul** the adsorbent.

This process is appropriate for relatively low concentrations of heavy hydrocarbons. It can also be appropriate if the gas is at a high pressure, close to the **cricondenbar**. In this case, refrigeration processes become ineffective and separation by adsorption may offer the only way to obtain the required specifications.

Membrane Separation Process

Refrigeration and **cryogenic plants** traditionally have been used for NGL recovery. These plants have high capital and operating costs. Moreover, they contain numerous rotating parts and are complicated to operate.

The membrane separation process offers a simple and low-cost solution for removal and recovery of heavy hydrocarbons from natural gas. The separation process is based on a **high-flux membrane** that selectively permeates heavy hydrocarbons compared to methane. These hydrocarbons permeate the membrane and are recovered as a liquid after recompression and condensation. The residue stream from the membrane is partially depleted of heavy hydrocarbons and is then sent to a sales gas stream.

Membrane systems are very versatile and are designed to process a wide range of feed conditions. With very compact footprint and low weight, these systems are well suited for offshore applications.

NGL Fractionation

The bottom liquid from the NGL recovery plant may be sold as a mixed product. This is common for small, isolated plants where there is insufficient local demand. The mixed product is transported by truck, rail, barge, or pipeline to a central location for further processing. Often it is more economical to **fractionate** the liquid into its various components, which have a market value as purity products. However, as the relative prices of natural gas and NGLs fluctuate, the relative incentive to extract the NGLs from the gas changes. The level of NGL extraction from natural gas is somewhat discretionary. Safety issues dictate the minimum extraction level, whereas a balance between the technology and the relative market value of the NGLs determines the maximum extraction level.

The process of separating a stream of NGLs into its components is called fractionation. At the fractionation plant, liquids will be separated into commercial quality products and then delivered to the market by tankers and tank trucks. NGLs are fractionated by heating mixed NGL streams and passing them through a series of **distillation towers**. Fractionation takes advantage of the differing boiling points of the various NGL products. As the temperature of the NGLstream is increased, the lightest (lowest boiling point) NGL product boils off the top of the tower as a gas where it is then condensed into a purity liquid that is routed to storage. The heavier liquid mixture at the bottom of the first tower is routed to the second tower where the process is repeated and a different NGL product is separated and stored. This process is repeated until the NGLs have been separated into their components.

Gasoline and LPG Treating

Natural gasoline (condensate) and LPG streams are often contaminated with acidic compounds. Especially objectionable are H_2S, mercaptans, and elemental sulfur. Natural gasoline containing H_2S has objectionable odor and is corrosive. Mercaptans give an objectionable odor to gasoline and elemental sulfur makes the gasoline corrosive.

H_2S in the LPG can result in the formation of free sulfur or mercaptans when used as feedstock. Mercaptans will impart strong odors to an LPG as well as its combustion products if they are present in any significant quantity. The presence of significant quantities of CO_2 can also cause problems by raising the vapor pressure of the LPG and lowering the heating value. COS and CS_2, although not corrosive in LPG, will **hydrolyze** slowly to H_2S in the presence of free water and cause the product to become corrosive.

H_2S and CO_2 can be removed from LPG and gasoline by liquid – liquid contacting processes using a **caustic solution**, aqueous alkanolamines, or solid KOH. When quantities of H_2S and CO_2 components are small, a simple **caustic wash** is both effective and economical. However, as the quantity of contaminants rises, the caustic supply and disposal costs render this approach impractical. Amine treating is a very attractive alternative, especially when there is already an amine gas treating unit on site.

Words and Expressions

condensate	凝析油
profitability	利润率
refrigeration	冷冻
adsorbent	吸附剂
foul	污染
cricondenbar	临界凝析压力
fractionate	分馏
hydrolyze	水解

Phrases and Expressions

natural gas liquid	天然气凝析液体
natural gasoline	天然汽油
gasoline – blending stock	调和汽油
shrinkage value	缩水价值
economic incentive	经济效益
retrograde condensation	反凝析
silica gel	硅胶
activated charcoal	活性炭
activated alumina	活性氧化铝
membrane separation process	薄膜分离法
cryogenic plant	低温装置
high – flux membrane	高流量的薄膜
distillation tower	蒸馏塔
caustic solution	碱液
caustic wash	碱洗

Language Focus

1. Recovery of NGL components in gas not only may be required for hydrocarbon dew point control in a natural gas stream, but also yields a source of revenue, as NGLs normally have significantly greater value as separate marketable products than as part of the natural gas stream.

(*参考译文:提取天然气凝析液不仅是为了控制天然气露点,而且 NGL 作为独立的市场产品比其在天然气中价值更高,可以获得更多收益。)

本句中" recovery of NGL components in gas "是主语,谓语是" not only... but also... "连接的两个并列结构" may be required for hydrocarbon dew point control in a natural gas stream "和" yields a source of revenue ",意为"不仅……而且……";第一个" as "作连词,意为"如同";后面的两个" as "作介词,意为"作为"。

2. The removal of natural gas liquids usually takes place in a relatively centralized processing plant, where the recovered NGLs are then treated to meet commercial specifications before moving into the NGL transportation infrastructure.

(参考译文：天然气液回收一般在比较集中的处理厂进行，回收的天然气液经过处理符合销售标准后，再进入天然气液输送设施。)

本句中"where"引导非限制性定语从句，相当于"in a relatively centralized processing plant"；"before"是连词，表示"在……之前"的时间概念。

3. As the temperature of the NGL stream is increased, the lightest (lowest boiling point) NGL product boils off the top of the tower as a gas where it is then condensed into a purity liquid that is routed to storage.

(参考译文：随着 NGL 流体温度的增加，最轻的 NGL 产品(沸点最低)在塔顶气化，然后冷凝成纯净的液体流到储存罐。)

本句中第一个"as"作连词，引导时间状语从句"the temperature of the NGL stream is increased"，意为"随着"；第二个"as"作介词，意为"作为"；"where"引导限制性定语从句，相当于"at the top of the tower"；"that"引导限制性定语从句，指代 "a purity liquid"。

Reinforced Learning

I. Answer the following questions for a comprehension of the text.

1. What are the advantages of recovery of NGL components?

2. What are NGL recovery processes?

3. What is the simple and low – cost solution that the membrane separation process offers?

4. How is fractionation conducted?

5. How to remove H_2S and CO_2 from LPG and gasoline?

II. Multiple choice: choose the correct one from the alternative answers to give the exact meaning of the words.

1. The price difference between selling NGL as a liquid and as fuel, commonly referred to as the "shrinkage value," often dictates the recovery level desired by the gas processors.

A. sinking wealth   B. contraction price

C. increased value   D. narrowed concept

2. Regardless of the economic incentive, however, gas usually must be processed to meet the specification for safe delivery and combustion.

A. inventiveness   B. initiative   C. benefit   D. crisis

3. Hence, NGL recovery profitability is not the only factor in determining

the degree of NGL extraction.

    A. payoff   B. improvement

    C. removal   D. specification

4. The main difference is that, in NGL absorption, an absorbing oil is used as opposed to glycol.

    A. in agreement with   B. in conformity to

    C. very similar to   D. in contrast to

5. They are expensive and more complex to operate, and it is difficult to predict their efficiency at removing liquids from the gas as the lean oil deteriorates with time.

    A. improves   B. decays

    C. ameliorates   D. convalesces

6. Activated alumina cannot be used in the presence of heavy hydrocarbons, which foul the adsorbent.

    A. dirty   B. adsorb   C. remove   D. construct

7. Refrigeration and cryogenic plants traditionally have been used for NGL recovery.

    A. operating   B. low – cost

    C. hypothermia   D. high – flux

8. The separation process is based on a high – flux membrane that selectively permeates heavy hydrocarbons compared to methane.

    A. rate   B. tech   C. capital   D. flow

9. Often it is more economical to fractionate the liquid into its various components, which have a market value as purity products.

    A. fracture   B. separate   C. function   D. obtain

10. The level of NGL extraction from natural gas is somewhat discretionary.

    A. random   B. selective   C. ineffective   D. appropriate

III. Multiple choice: read the four suggested translations and choose the best answer.

1. These heavier hydrocarbon liquids, commonly referred to as natural gas liquids (NGLs), include ethane, propane, butanes, and natural gasoline (condensate).

    A. 天然汽油   B. 天然气液   C. 调和汽油   D. 天然气露点

2. Lighter NGL fractions can be sold as fuel or feedstock to refineries and petrochemical plants, while the heavier portion can be used as <u>gasoline - blending stock</u>.

    A. 凝析油　　　B. 调和汽油　C. 天然汽油　D. 液化石油气

3. Obviously any cooling outside the <u>retrograde condensation zone</u> will induce condensation and yield NGL.

    A. 制冷工艺专区　　　　B. 机械制冷带

    C. 天然气冷凝区　　　　D. 反凝析区域

4. It is therefore important to know where we are on the <u>phase envelope</u>.

    A. 阶段信封　　　B. 位相包层　C. 相包络线　D. 相位包膜

5. <u>Mechanical refrigeration</u> is the simplest and most direct process for NGL recovery.

    A. 机械制冷法　　B. 制冷工艺　C. 冷却气体　D. 机械冷冻

6. Note that an oil absorption plant cannot recover ethane and propane effectively, where it requires circulating large amounts of <u>absorption oil</u>, demands attendant maintenance, and consumes too much fuel.

    A. 脱吸油　　　B. 凝析油　　C. 吸附油　　D. 调和油

7. The adsorbent may be <u>silica gel</u> or activated charcoal.

    A. 硅胶　　　B. 活性炭　　C. 重烃　　　D. 吸附剂

8. It can also be appropriate if the gas is at a high pressure, close to the <u>cricondenbar</u>.

    A. 贫油吸附装置　　　　B. 蒸馏塔

    C. 胺处理装置　　　　D. 临界凝析压力

9. Especially <u>objectionable</u> are H_2S, mercaptans, and elemental sulfur.

    A. 重要的　　　B. 适合的　　C. 严重的　　D. 客观的

10. H_2S in the LPG can result in the formation of free sulfur or mercaptans when used as <u>feedstock</u>.

    A. 备选　　　B. 原料　　　C. 存货　　　D. 备份

IV. Put the following sentences into Chinese.

1. The price difference between selling NGL as a liquid and as fuel, commonly referred to as the "shrinkage value," often dictates the recovery level desired by the gas processors.

2. Regardless of the economic incentive, however, gas usually must be processed to meet the specification for safe delivery and combustion.

3. This method uses adsorbents that have the capability to adsorb heavy hydrocarbons from natural gas.

4. The heavier liquid mixture at the bottom of the first tower is routed to the second tower where the process is repeated and a different NGL product is separated and stored.

5. The removal of natural gas liquids usually takes place in a relatively centralized processing plant, where the recovered NGLs are then treated to meet commercial specifications before moving into the NGL transportation infrastructure.

V. Put the following paragraphs into Chinese.

1. Lean oil absorption is the oldest and least efficient process to recover NGLs. Note that an oil absorption plant cannot recover ethane and propane effectively, where it requires circulating large amounts of absorption oil, demands attendant maintenance, and consumes too much fuel. Lean oil absorption plants are not as popular as they once were and are rarely, if ever, constructed anymore. They are expensive and more complex to operate, and it is difficult to predict their efficiency at removing liquids from the gas as the lean oil deteriorates with time.

2. The bottom liquid from the NGL recovery plant may be sold as a mixed product. This is common for small, isolated plants where there is insufficient local demand. The mixed product is transported by truck, rail, barge, or pipeline to a central location for further processing. Often it is more economical to fractionate the liquid into its various components, which have a market value as purity products. However, as the relative prices of natural gas and NGLs fluctuate, the relative incentive to extract the NGLs from the gas changes. The level of NGL extraction from natural gas is somewhat discretionary. Safety issues dictate the minimum extraction level, whereas a balance between the technology and the relative market value of the NGLs determines the maximum extraction level.

Chapter 5   Acid Natural Gas Treating

5.1   Acid Gas Treating(1)

⌘ Guidance to Reading

The term "acid gas" in this chapter is referred to the natural gas contai-ning significant quantities of hydrogen sulfide (H_2S) and carbon dioxide (CO_2), or similar acidic gases. The terms "acid gas" and "sour gas" are of-ten incorrectly treated as synonyms. Strictly speaking, a sour gas is any gas that specifically contains hydrogen sulfide in significant amounts; the natural gas free from hydrogen sulfide is referred to as "sweet"; and an acid gas is any gas that contains significant amounts of acidic gases such as carbon dioxide or hydrogen sulfide. Thus, carbon dioxide by itself is an acid gas but not a sour gas.

⌘ Text

Natural gas, while **ostensibly** being hydrocarbon in nature, contains large amounts of acid gases, such as hydrogen sulfide and carbon dioxide. Natural gas containing hydrogen sulfide or carbon dioxide is referred to as *sour*, and natural gas free from hydrogen sulfide is referred to as *sweet*. Because of the **corrosiveness** nature of hydrogen sulfide and carbon dioxide in the presence of wa-ter (giving rise to an acidic **aqueous solution**) and because of the **toxicity** of hydrogen sulfide and the lack of heating value of carbon dioxide, natural gas being prepared for sales is required to contain no more than 5 ppm hydrogen sulfide and to have a heating value of no less than 920 to 980 Btu/scf. The ac-tual specifications depend on the use, the country where the gas is used, and the contract. However, because natural gas has a wide range of composition, including the concentration of the two acid gases, processes for the removal of acid gases vary and are subject to choice based on the desired end product.

There are many **variables** in treating natural gas. The precise area of appli-cation of a given process is difficult to define. Several factors must be considered: (1) types and concentrations of contaminants in the gas, (2) the degree of contaminant removal desired, (3) the **selectivity** of acid gas removal required, (4) the temperature, pressure, volume, and composition of the gas to be processed, (5) the carbon dioxide – hydrogen sulfide ratio in the gas, and (6) the **desirability** of sulfur recovery due to process economics or environmental issues.

In addition to hydrogen sulfide and carbon dioxide, gas may contain other contaminants, such as **mercaptans** and **carbonyl sulfide**. The presence of these impurities may eliminate some of the sweetening processes, as some processes remove large amounts of acid gas but not to a sufficiently low concentration. However, there are those processes that are not designed to remove (or are incapable of removing) large amounts of acid gases. These processes are also capable of removing the acid gas impurities to very low levels when the acid gases are there in low to medium concentrations in the gas.

Process selectivity indicates the preference with which the process removes one acid gas component relative to (or in preference to) another. For example, some processes remove both hydrogen sulfide and carbon dioxide; other processes are designed to remove hydrogen sulfide only. It is important to consider the process selectivity for, say, hydrogen sulfide removal compared to carbon dioxide removal that ensures minimal concentrations of these components in the product, thus the need for consideration of the carbon dioxide to hydrogen sulfide in the gas stream.

Acid Gas Removal Processes

The processes that have been developed to accomplish gas purification vary from a simple once – through wash operation to complex multistep recycling systems. In many cases, process complexities arise because of the need for recovery of the materials used to remove the contaminants or even recovery of the contaminants in the original, or altered, form.

There are two general processes used for acid gas removal: adsorption and absorption. Adsorption is a physical – chemical phenomenon in which the gas is concentrated on the surface of a solid or liquid to remove impurities. Usually, carbon is the adsorbing medium, which can be regenerated upon desorption. The quantity of material adsorbed is proportional to the surface area of the solid and, consequently, **adsorbents** are usually **granular solids** with a large surface area per unit mass. Subsequently, the captured gas can be desorbed with hot air or steam either for recovery or for thermal destruction. Adsorbers are widely used to increase a low gas concentration prior to **incineration** unless the gas concentration is very high in the inlet air stream. Adsorption is also employed to reduce problem odors from gases. There are several limitations to the use of adsorption systems, but it is generally felt that the major one is the requirement for minimization of **particulate matter** and/or condensation of liquids (e. g. , water vapor) that could mask the adsorption surface and reduce its efficiency drastically.

Absorption differs from adsorption in that it is not a physical – chemical surface phenomenon, but an approach in which the absorbed gas is ultimately distributed throughout the absorbent (liquid). The process depends only on physical solubility and may include chemical reactions in the liquid phase (chemisorption). Common absorbing media used are water, **aqueous amine solutions**, **sodium carbonate**, and **nonvolatile hydrocarbon oils**, depending on the type of gas to be absorbed. Usually, the gas – liquid contactor designs that are employed are **plate columns** or **packed beds**.

Absorption is achieved by dissolution (a physical phenomenon) or by reaction (a chemical phenomenon). Chemical adsorption processes adsorb sulfur dioxide onto a carbon surface where it is oxidized (by oxygen in the flue gas) and absorbs moisture to give sulfuric acid **impregnated** into and on the adsorbent.

As currently practiced, acid gas removal processes involve the chemical reaction of the acid gases with a solid oxide (such as iron oxide) or selective absorption of the contaminants into a liquid (such as **ethanolamine**) that is passed countercurrent to the gas. Then the absorbent is stripped of the gas components (regeneration) and recycled to the absorber. The process design will vary and, in practice, may employ multiple absorption columns and multiple regeneration columns.

Liquid absorption processes [which usually employ temperatures below 50℃ (120 ℉)] are classified either as **physical solvent processes** or as **chemical solvent processes**. The former processes employ an **organic solvent**, low temperatures, or high pressure. In chemical solvent processes, absorption of the acid gases is achieved mainly by use of **alkaline solutions** such as **amines** or

carbonates. Regeneration (desorption) can be brought about by the use of reduced pressures and/or high temperatures, whereby the acid gases are stripped from the solvent.

Amine washing of natural gas involves chemical reaction of the amine with any acid gases with the liberation of an appreciable amount of heat and it is necessary to compensate for the absorption of heat. **Amine derivatives** such as ethanolamine (monoethanolamine), diethanolamine, triethanolamine, **methyldiethanolamine**, **diisopropanolamine**, and **diglycolamine** have been used in commercial applications.

Words and Expressions

ostensibly	表面上
corrosiveness	腐蚀性
toxicity	毒性
variable	变量
selectivity	选择性
desirability	意愿
mercaptan	硫醇
adsorbent	吸附剂
incineration	煅烧
nonvolatile	非挥发性的
impregnate	浸入
ethanolamine	乙醇胺
amine	醇胺类
carbonate	碳酸盐类
methyldiethanolamine	甲基二乙醇胺
diisopropanolamine	二异丙醇胺
diglycolamine	二甘醇胺

Phrases and Expressions

aqueous solution	水溶液
carbonyl sulfide	羰基硫化物
granular solid	颗粒状固体
particulate matter	悬浮微粒

aqueous amine solution 　　　醇胺水溶液

sodium carbonate 　　　碳酸钠

hydrocarbon oil 　　　烃油

plate column 　　　板式塔

packed bed 　　　填料塔

physical solvent process 　　　物理溶剂法

chemical solvent process 　　　化学溶剂法

organic solvent 　　　有机溶剂

alkaline solution 　　　碱溶液

amine derivative 　　　胺衍生物

🔲 Language Focus

1. Because of the corrosiveness nature of hydrogen sulfide and carbon dioxide in the presence of water (giving rise to an acidic aqueous solution) and because of the toxicity of hydrogen sulfide and the lack of heating value of carbon dioxide, natural gas being prepared for sales is required to contain no more than 5 ppm hydrogen sulfide and to have a heating value of no less than 920 to 980 Btu/scf.

（参考译文：一旦有水，硫化氢和二氧化碳就会产生腐蚀性（形成一种酸性水溶液），而且硫化氢有毒，二氧化碳没有热值，因此，待售天然气的硫化氢含量须不高于 $5mL/m^3$,并且热值不低于 $920 \sim 980\ Btu/ft^3$ 。）

本句主谓结构是"natural gas being prepared for sales is required to..."; "Because of the corrosiveness nature of hydrogen sulfide and carbon dioxide in the presence of water (giving rise to an acidic aqueous solution) and because of the toxicity of hydrogen sulfide and the lack of heating value of carbon dioxide" 是由两个 "because of"引导的原因状语。

2. It is important to consider the process selectivity for, say, hydrogen sulfide removal compared to carbon dioxide removal that ensures minimal concentrations of these components in the product, thus the need for consideration of the carbon dioxide to hydrogen sulfide in the gas stream.

（参考译文："考虑工艺选择十分重要，比如硫化氢脱除法和二氧化碳脱除法相比，二氧化碳脱除法能保证产品中这些组分含量最低，因此就要考虑选择脱除气体中的二氧化碳而不是硫化氢。）

本句中"it"为形式主语，真正的主语是"to consider the process selectivity

for";"say"意为"比如",后面是举例;"compared to carbon dioxide removal"
为过去分词作后置定语,修饰"hydrogen sulfide removal";"that"引导限制性
定语从句,指代 "carbon dioxide removal"。

Reinforced Learning

I. Answer the following questions for a comprehension of the text.

1. What are variables in treating natural gas?

2. What is process selectivity?

3. What are the two general processes used for acid gas removal?

4. What are current acid gas removal processes?

5. What amine derivatives have been used in commercial applications?

II. Multiple choice: choose the correct one from the alternative answers to give the exact meaning of the words.

1. Natural gas, while <u>ostensibly</u> being hydrocarbon in nature, contains large amounts of acid gases, such as hydrogen sulfide and carbon dioxide.

    A. apparently   B. facially   C. obviously   D. naturally

2. The corrosiveness nature of hydrogen sulfide and carbon dioxide in the presence of water (<u>giving rise to</u> an acidic aqueous solution) and because of the toxicity of hydrogen sulfide and the lack of heating value of carbon dioxide, natural gas being prepared for sales is required to contain no more than 5 ppm hydrogen sulfide and to have a heating value of no less than 920 to 980 Btu/scf.

    A. leading to   B. resulting from

    C. according to   D. rising from

3. However, because natural gas has a wide range of composition, including the concentration of the two acid gases, processes for the removal of acid gases vary and <u>are subject to</u> choice based on the desired end product.

    A. are made of   B. are made from

    C. are involved by   D. are decided by

4. Process selectivity indicates the preference with which the process removes one acid gas component relative to (or <u>in preference to</u>) another.

    A. inferior to   B. in conformity to

    C. superior to   D. in agreement to

5. It is important to consider the process selectivity for, say, hydrogen sulfide removal compared to carbon dioxide removal that ensures minimal concentrations of these components in the product, thus the need for consideration of the carbon dioxide to hydrogen sulfide in the gas stream.

A. speak   B. for example

C. in particular   D. not to mention

6. Adsorbers are widely used to increase a low gas concentration prior to incineration unless the gas concentration is very high in the inlet air stream.

A. dissolution   B. chemisorption

C. burning   D. condensation

7. There are several limitations to the use of adsorption systems, but it is generally felt that the major one is the requirement for minimization of particulate matter and/or condensation of liquids (e. g. , water vapor) that could mask the adsorption surface and reduce its efficiency drastically.

A. originally   B. usually

C. proportionally   D. completely

8. Chemical adsorption processes adsorb sulfur dioxide onto a carbon surface where it is oxidized (by oxygen in the flue gas) and absorbs moisture to give sulfuric acid impregnated into and on the adsorbent.

A. absorbed   B. infused

C. achieved   D. desorbed

9. As currently practiced, acid gas removal processes involve the chemical reaction of the acid gases with a solid oxide (such as iron oxide) or selective absorption of the contaminants into a liquid (such as ethanolamine) that is passed countercurrent to the gas.

A. through   B. reflux   C. crossing   D. back

10. Amine washing of natural gas involves chemical reaction of the amine with any acid gases with the liberation of an appreciable amount of heat and it is necessary to compensate for the absorption of heat.

A. substantial   B. measurable

C. unpredictable   D. incredible

Ⅲ. Multiple choice：read the four suggested translations and choose the best answer.

1. Natural gas containing hydrogen sulfide or carbon dioxide is referred to

as *sour*, and natural gas free from hydrogen sulfide is referred to as *sweet*.

    A. 硫醇或羰基硫化物   B. 硫化氢或二氧化碳

    C. 水或醇胺水溶液   D. 氢氧化钠或碳酸钠

2. There are many <u>variables</u> in treating natural gas.

    A. 量化   B. 变数   C. 变化   D. 变量

3. The processes that have been developed to accomplish gas purification vary from a simple <u>once – through</u> wash operation to complex <u>multistep</u> recycling systems.

    A. 单程 多程   B. 直通 多次

    C. 一次 多步   D. 一次 多个

4. There are two general processes used for acid gas removal: <u>adsorption and absorption</u>.

    A. 吸着再吸着   B. 吸附后吸收

    C. 吸收作用和吸附作用   D. 吸附作用和吸收作用

5. Usually, carbon is the adsorbing medium, which can be regenerated upon <u>desorption</u>.

    A. 吸收   B. 吸取   C. 解吸   D. 解吐

6. The quantity of material adsorbed is proportional to the surface area of the solid and, consequently, adsorbents are usually <u>granular solids</u> with a large surface area per unit mass.

    A. 固体颗粒   B. 粒状固体

    C. 颗粒物质   D. 固体吸附剂

7. Common absorbing media used are water, <u>aqueous amine solutions</u>, caustic, sodium carbonate, and nonvolatile hydrocarbon oils, depending on the type of gas to be absorbed.

    A. 氢氧化钠   B. 醇胺水溶液

    C. 碳酸钠溶液   D. 羰基硫化物

8. Usually, the gas – liquid contactor designs that are employed are <u>plate columns</u> or packed beds.

    A. 板式塔   B. 平底柱

    C. 气液发生器   D. 循环系统

9. The process design will vary and, in practice, may employ <u>multiple absorption columns and multiple regeneration columns</u>.

A. 多次汲取器和多次生长器　　　B. 多元吸入器和多元重生器

C. 多重吸收器和多重再生器　　　D. 多级吸收器和多级再生器

10. Amine <u>derivatives</u> such as ethanolamine (monoethanolamine), dietha-nolamine, triethanolamine, methyldiethanolamine, diisopropanolamine, and dig-lycolamine have been used in commercial applications.

A. 导出物　　　　B. 衍生品　　　　C. 派生物　　　　D. 衍生物

Ⅳ. Put the following sentences into Chinese.

1. The presence of these impurities may eliminate some of the sweetening processes, as some processes remove large amounts of acid gas but not to a suf-ficiently low concentration.

2. In many cases, process complexities arise because of the need for re-covery of the materials used to remove the contaminants or even recovery of the contaminants in the original, or altered, form.

3. Chemical adsorption processes adsorb sulfur dioxide onto a carbon sur-face where it is oxidized (by oxygen in the flue gas) and absorbs moisture to give sulfuric acid impregnated into and on the adsorbent.

4. As currently practiced, acid gas removal processes involve the chemical reaction of the acid gases with a solid oxide (such as iron oxide) or selective absorption of the contaminants into a liquid (such as ethanolamine) that is pas-sed countercurrent to the gas.

5. Amine washing of natural gas involves chemical reaction of the amine with any acid gases with the liberation of an appreciable amount of heat and it is necessary to compensate for the absorption of heat.

Ⅴ. Put the following paragraphs into Chinese.

1. Natural gas, while ostensibly being hydrocarbon in nature, contains large amounts of acid gases, such as hydrogen sulfide and carbon dioxide. Nat-ural gas containing hydrogen sulfide or carbon dioxide is referred to as *sour*, and natural gas free from hydrogen sulfide is referred to as *sweet*. Because of the corrosiveness nature of hydrogen sulfide and carbon dioxide in the presence of water and because of the toxicity of hydrogen sulfide and the lack of heating value of carbon dioxide, natural gas being prepared for sales is requried to con-tain no more than 5 ppm hydrogen sulfide and to have a heating value of no less than 920 to 980 Btu/scf. The actual specifications of natural gas being prepared

for sales depend on the use, the country where the gas is used, and the contract. However, because natural gas has a wide range of composition, including the concentration of the two acid gases, processes for the removal of acid gases vary and are subject to choice based on the desired end product.

2. Adsorption is a physical – chemical phenomenon in which the gas is concentrated on the surface of a solid or liquid to remove impurities. Usually, carbon is the adsorbing medium, which can be regenerated upon desorption. The quantity of material adsorbed is proportional to the surface area of the solid and, consequently, adsorbents are usually granular solids with a large surface area per unit mass. Subsequently, the captured gas can be desorbed with hot air or steam either for recovery or for thermal destruction. Absorption differs from adsorption in that it is not a physical – chemical surface phenomenon, but an approach in which the absorbed gas is ultimately distributed throughout the absorbent (liquid). The process depends only on physical solubility and may include chemical reactions in the liquid phase (chemisorption).

5.2　Acid Gas Treating(2)

Guidance to Reading

*Before a raw natural gas containing **hydrogen sulfide** and/ or carbon dioxide can be used, the raw gas must be treated to reduce impurities to acceptable levels. There are many variables in treating natural gas. The precise area of application of a given process is difficult to define for several governing factors. In addition to the two general processes (in the previous sec tion) used for acid gas removal: adsorption and absorption, other processes will be involved in this sec tion as batchtype process, amine process and **carbonate washing** and water washing.*

Text

Batchtype Type Processes

The most common type of process for acid gas removal is the **batchtype** process and may involve a chemical process in which the acid gas reacts chemically with the **cleaning agent**, usually a **metal oxide**. These processes are not merely physical separation processes in which the acid gas is removed by a physical phenomenon, such as adsorption. Thus, the batchtype processes have the common requirement that the process be operated as a batch system where, at the end of the cycle, the chemical agent must be changed or regenerated in order to continue treating. Batch processes are limited to removing small amounts of sulfur, i. e. , low gas flow rates and/or small concentrations of hydrogen sulfide.

These processes **scavenge** hydrogen sulfide and organic sulfur compounds (**mercaptans**) from gas streams through reactions with solid – based media. They are typically nonregenerable, although some are partially regenerable, losing activity upon each regeneration cycle. Most dry **sorption** processes are governed by the reaction of a metal oxide with H_2S to form a metal sulfide compound. For regenerable reactions, the metal sulfide compound can then react with oxygen to produce elemental sulfur and a regenerated metal oxide. The primary metal oxides used for dry sorption processes are **iron oxide** and **zinc oxide**.

Amine Processes

Chemical absorption processes with **aqueous alkanolamine solutions** are used for treating gas streams containing hydrogen sulfide and carbon dioxide. However, depending on the composition and operating conditions of the feed gas, different amines can be selected to meet the product gas **specification**. Amines are categorized as being primary, secondary, and tertiary depending on the degree of substitution of the central nitrogen by organic groups. Primary amines react directly with H_2S, CO_2, and carbonyl sulfide (COS). Examples of primary amines include **monoethanolamine** (**MEA**) and the **proprietary diglycolamine agent** (**DGA**). Secondry amines react directly with H_2S and CO_2 and react directly with some COS. The most common sec ondry amine is **diethanolamine** (**DEA**). Tertiary amines react directly with H_2S, react indirectly with CO_2, and react indirectly with little COS. The most common examples of tertiary amines are **methyldiethanolamine** (**MDEA**) and activated methyldiethanolamine.

Processes using ethanolamine and **potassium phosphate** are now widely used. The ethanolamine process, known as the **Girbotol process**, removes acid gases (hydrogen sulfide and carbon dioxide) from liquid hydrocarbons as well as from natural and from refinery gases. Depending on the application, special solutions such as mixtures of amines; amines with physical solvents, such as **sulfolane** and **piperazine**; and amines that have been partially neutralized with an acid such as **phosphoric acid** may also be used.

MEA is a stable compound and, in the absence of other chemicals, suffers no degradation or decomposition at temperatures up to its normal boiling point. DEA is a weaker base than MEA and therefore the DEA system does not typically suffer the same corrosion problems but does react with hydrogen sulfide and carbon dioxide. DEA also removes carbonyl sulfide and carbon dislduflide partially as its regenerable compound with COS and CS_2 without much solution losses.

One key difference among the various specialty amines is selectivity toward hydrogen sulfide. Instead of removing both hydrogen sulfide and carbon dioxide, as generic amines such as MEA and DEA do, some products readily remove hydrogen sulfide to specifications, but allow controlled amounts of carbon dioxide to slip through.

Carbonate Washing and Water Washing

Carbonate washing is a mild alkali process for emission control by the removal of acid gases (such as carbon dioxide and hydrogen sulfide) from gas streams and uses the principle that the rate of absorption of carbon dioxide by **potassium carbonate** increases with temperature. It has been demonstrated that the process works best near the temperature of reversibility of the reactions.

Water washing, in terms of the outcome, is ana log ous to washing with potassium carbonate, and it is also possible to carry out the desorption step by pressure reduction. The absorption is purely physical and there is also a relatively high absorption of hydrocarbons, which are liberated at the same time as the acid gases.

Methanol Based Processes

Methanol is probably one of the most versatile solvents in the natural gas proces sing industry. Historically, methanol was the first commercial organic physical solvent and has been used for hydrate inhibition, dehydration, gas sweetening, and liquids recovery. Most of these applications involve low temperature where the physical properties of methanol are advantageous compared with other solvents that exhibit high vis cos ity problems or even solids formation. Operation at low temperatures tends to suppress the most significant disadvantage of methanol, high solvent loss. Furthermore, methanol is relatively inexpensive and easy to produce, making the solvent a very attractive alternate for gas proces sing applications.

Methanol has favorable physical properties relative to other solvents except for vapor pressure. The benefits of the low viscosity of methanol at low temperature are manifested in the pressure drop improvement in the cold box of injection facilities and improved heat transfer. Methanol has a much lower surface tension relative to the other solvents. High surface tension tends to promote foaming problems in contactors. Methanol processes are probably not susceptible to foaming. However, the primary drawback of methanol is the high vapor pressure, which is several times greater than that of glycols or amines. To minimize methanol losses and enhance water and acid gas absorption, the absorber or separator temperatures are usually less than $-20\,°F$.

The high vapor pressure of methanol may initially appear to be a significant drawback because of high solvent losses. However, the high vapor pressure also has significant advantages. Although often not considered, lack of thorough mixing of the gas and solvent can pose significant problems. Because of the high vapor pressure, methanol is completely mixed in the gas stream before the cold box. Glycols, because they do not completely vaporize, may require special nozzles and nozzle placement in the cold box to prevent freeze – up. Solvent carryover to other downstream processes may also represent a significant problem. Because methanol is more volatile than glycols, amines, and other physical solvents, including lean oil, methanol is usually rejected in the regeneration step of these downstream processes. The stripper concentrates the methanol in the overhead condenser where it can be removed and further purified. Unfortunately, if glycols are carried over to amine units, the glycol becomes concentrated in the solution and potentially starts to degrade and possibly dilute the amine solution.

Words and Expressions

batchtype	间歇式
scavenge	清除
mercaptan	硫醇
sorption	吸附

specification	规格
monoethanolamine	一乙醇胺
proprietary	专用的
diethanolamine	二乙醇胺
methyldiethanolamine	甲基二乙醇胺
sulfolane	环丁砜
piperazine	六氢吡嗪

Phrases and Expressions

hydrogen sulfide	硫化氢
carbonate washing	碱洗
cleaning agent	清洗剂
metal oxide	金属氧化物
iron oxide	氧化铁
zinc oxide	氧化锌
aqueous alkanolamine solution	醇胺水溶液
diglycolamine agent	二乙二醇胺
potassium phosphate	磷酸钾
girbotol process	乙醇胺法
phosphoric acid	磷酸
potassium carbonate	碳酸钾

Language Focus

1. The most common type of process for acid gas removal is the batchtype process and may involve a chemical process in which the acid gas reacts chemically with the cleaning agent, usually a metal oxide.

（参考译文：最常见的酸性气体脱除方法是间歇处理法,其中一般涉及酸性气体与清洗剂发生化学反应,清洗剂一般为金属氧化物。）

本句中"the most common type of process for acid gas removal"是主语,"and"连接两个并列谓语"is the batchtype process"和"may involve a chemical process in which the acid gas reacts chemically with the cleaning agent, usually a metal oxide";"in which"引导限制性定语从句,相当于"in a chemical process";"usually a metal oxide"是"the cleaning agent"的同位语。

2. Thus, the batch – type processes have the common requirement that the

process be operated as a batch system where, at the end of the cycle, the chemical agent must be changed or regenerated in order to continue treating.

（参考译文：间歇处理的各个流程没有特殊的技术要求，在循环终端对化学处理剂进行更换或再生利用。）

本句主句为"the batch – type processes have the common requirement"；"that"引导同位语从句，说明 requirement 的内容，做 requirement 的同位语时，从句用虚拟语气，此句"be"动词前省略了"should"；"where"引导限制性定语从句，相当于"in a batch system"。

3. Instead of removing both hydrogen sulfide and carbon dioxide, as generic amines such as MEA and DEA do, some products readily remove hydrogen sulfide to specifications, but allow controlled amounts of carbon dioxide to slip through.

（参考译文：某些胺类可以将硫化氢脱除到符合标准要求，并允许一定量的二氧化碳通过，不像一乙醇胺和二乙醇胺那样既清除硫化氢又清除二氧化碳。）

本句中"instead of"作介词短语，意为"代替"；第一个"as"作连词，意为"正如"；"but"作连词，意为"但是"。

Reinforced Learning

Ⅰ. Answer the following questions for a comprehension of the text.

1. What is the most common type of process for acid gas removal?

2. What are used for treating gas streams containing hydrogen sulfide and carbon dioxide?

3. What is the key difference among the various specialty amines?

4. What is carbonate washing?

5. What are applications of Methanol?

Ⅱ. Multiple choice: choose the correct one from the alternative answers to give the exact meaning of the words.

1. The most common type of process for acid gas removal is the batchtype process and may involve a chemical process in which the acid gas reacts chemically with the cleaning agent, usually a metal oxide.

A. bunch   B. internal   C. interval   D. terminal

2. Thus, the batch – type processes have the common requirement that the

process be operated as a batch system where, at the end of the cycle, the chemical agent must be changed or regenerated in order to continue treating.

    A. practice   B. necessity   C. conduct   D. demand

    3. These processes <u>scavenge</u> hydrogen sulfide and organic sulfur compounds (mercaptans) from gas streams through reactions with solid – based media.

    A. clear   B. scatter   C. contain   D. manage

    4. They are typically <u>nonregenerable</u>, although some are partially regenerable, losing activity upon each regeneration cycle.

    A. productive   B. renewable

    C. nonrecyclable   D. infertile

    5. <u>Processes</u> using ethanolamine and potassium phosphate are now widely used.

    A. Procedures   B. Treatments

    C. Stages   D. Spheres

    6. Depending on the application, special solutions such as mixtures of amines; amines with physical solvents, such as sulfolane and piperazine; and amines that have been partially <u>neutralized</u> with an acid such as phosphoric acid may also be used.

    A. treated   B. operated

    C. processed   D. counteracted

    7. DEA also removes carbonyl sulfide and carbon dislduflide partially as its regenerable compound with COS and CS_2 without much solution <u>losses</u>.

    A. deprivation   B. profit

    C. discovery   D. compression

    8. It has been demonstrated that the process works best near the temperature of <u>reversibility</u> of the reactions.

    A. review   B. revision   C. interval   D. invertibility

    9. Historically, methanol was the first <u>commercial</u> organic physical solvent and has been used for hydrate inhibition, dehydration, gas sweetening, and liquids recovery.

    A. mercantile   B. official   C. legal   D. public

    10. However, the primary <u>drawback</u> of methanol is the high vapor pressure, which is several times greater than that of glycols or amines.

A. withdrawal　　　　　　B. disadvantage

C. circumstance　　　　　D. requirement

Ⅲ. Multiple choice：read the four suggested translations and choose the best answer.

1. Batch processes are limited to removing small amounts of sulfur, i. e. , <u>low gas flow rates</u> and/or small concentrations of hydrogen sulfide.

A. 天然气低流速　　　　B. 低硫天然气流速

C. 少量天然气流速　　　D. 天然气低硫速度

2. These processes scavenge hydrogen sulfide and <u>organic sulfur compounds</u>（mercaptans）from gas streams through reactions with solid – based media.

A. 金属氧化物　B. 有机硫化物　C. 单质硫　　　D. 硫化氢

3. Most <u>dry sorption processes</u> are governed by the reaction of a metal oxide with H₂S to form a metal sulfide compound.

A. 干式吸附工艺　　　　B. 干燥吸着过程

C. 干吸收处理　　　　　D. 干吸剂

4. The primary metal oxides used for dry sorption processes are iron oxide and <u>zinc oxide</u>.

A. 氧化铁　　　　　　　B. 金属硫化物

C. 氧化锌　　　　　　　D. 金属氧化物

5. Chemical absorption processes with <u>aqueous alkanolamine solutions</u> are used for treating gas streams containing hydrogen sulfide and carbon dioxide.

A. 一乙醇胺　　　　　　B. 二乙醇胺

C. 二乙二醇胺　　　　　D. 醇胺水溶液

6. However, depending on the composition and operating conditions of <u>the feed gas</u>, different amines can be selected to meet the product gas specification.

A. 喂养气　　B. 流入气　　C. 原料气　　　D. 进口气

7. Amines are categorized as being <u>primary, sec ondary, and tertiary</u> depending on the degree of substitution of the central nitrogen by organic groups.

A. 第一、第二、第三　　　B. 初级、中级、高级

C. 首胺、次胺、再胺　　　D. 伯胺、仲胺、叔胺

8. Examples of primary amines include <u>monoethanolamine（MEA）</u> and the <u>proprietary diglycolamine agent（DGA）</u>.

A. 二氧化碳 硫化氢

B. 一乙醇胺 专用二乙二醇胺

C. 羰基硫 二乙醇胺

D. 乙醇胺 磷酸钾

9. Carbonate washing is a mild alkali process for emission control by the removal of acid gases (such as carbon dioxide and hydrogen sulfide) from gas streams and uses the principle that the rate of absorption of carbon dioxide by potassium carbonate increases with temperature.

A. 碳酸钾   B. 碱洗   C. 硫化氢   D. 乙醇胺

10. Unfortunately, if glycols are carried over to amine units, the glycol becomes concentrated in the solution and potentially starts to degrade and possibly dilute the amine solution.

A. 碳酸钾洗涤设备   B. 醇胺装置

C. 甲醇蒸汽压力器   D. 胺脱硫单位

Ⅳ. **Put the following sentences into Chinese.**

1. For regenerable reactions, the metal sulfide compound can then react with oxygen to produce elemental sulfur and a regenerated metal oxide.

2. Tertiary amines react directly with H_2S, react indirectly with CO_2, and react indirectly with little COS.

3. The ethanolamine process, known as the Girbotol process, removes acid gases (hydrogen sulfide and carbon dioxide) from liquid hydrocarbons as well as from natural and from refinery gases.

4. Carbonate washing is a mild alkali process for emission control by the removal of acid gases from gas streams and uses the principle that the rate of absorption of carbon dioxide by potassium carbonate increases with temperature.

5. To minimize methanol losses and enhance water and acid gas absorption, the absorber or separator temperatures are usually less than $-20\,°F$.

Ⅴ. **Put the following paragraphs into Chinese.**

1. The most common type of process for acid gas removal is the batchtype process and may involve a chemical process in which the acid gas reacts chemically with the cleaning agent, usually a metal oxide. These processes are not merely physical separation processes in which the acid gas is removed by a

physical phenomenon, such as adsorption. Thus, the batch – type processes have the common requirement that the process be operated as a batch system where, at the end of the cycle, the chemical agent must be changed or regenerated in order to continue treating. Batch processes are limited to removing small amounts of sulfur, i. e. , low gas flow rates and/or small concentrations of hydrogen sulfide.

2. The high vapor pressure of methanol may initially appear to be a significant drawback because of high solvent losses. However, the high vapor pressure also has significant advantages. Although often not considered, lack of thorough mixing of the gas and solvent can pose significant problems. Because of the high vapor pressure, methanol is completely mixed in the gas stream before the cold box. Glycols, because they do not completely vaporize, may require special nozzles and nozzle placement in the cold box to prevent freeze – up.

5.3　Acid Gas Treating(3)

Guidance to Reading

The side stream from acid gas treating units consists mainly of hydrogen sulfide/or carbon dioxide. Carbon dioxide is usually vented to the atmosphere or recovered for CO_2 floods. However, the release of H_2S to the atmosphere may be limited by environmental regulations. Sulfur recovery refers to the conversion of hydrogen sulfide (H_2S) to elemental sulfur. The most common conversion method used is the Claus process. Approximately 90 to 95 percent of recovered sulfur is produced by the Claus process. First patented in 1883 by the scientist Carl Friedrich Claus, the Claus process has become the industry standard.

Text

Other Processes

The process using **potassium phosphate** is used in the same way as the Girbotol process to remove acid gases from liquid hydrocarbons as well as from gas streams. The treatment solution is a water solution of **tripotassium phosphate** (K_3PO_4), which is circulated through an **absorber tower** and a **reactivator tower** in much the same way as the ethanolamine is circulated in the Girbotol process; the solution is regenerated thermally.

Other processes include the **Alkazid process.** The hot **potassium carbonate** process decreases the acid content of natural and refinery gas from as much as 50% to as low as 0.5% and operates in a unit similar to that used for amine treating.

The **Giammarco – Vetrocoke** process is used for hydrogen sulfide and/or carbon dioxide removal. In the hydrogen sulfide removal sec tion, the **reagent** consists of **sodium** or **potassium** carbonates containing a mixture of **arsenites** and **arsenates**; the carbon dioxide removal sec tion utilizes hot aqueous alkali carbonate solution activated by **arsenic trioxide** or **selenous acid** or **tellurous acid**.

Molecular **sieves** are highly selective for the removal of hydrogen sulfide (as well as other sulfur compounds) from gas streams and over a continuously high absorption efficiency. They are also an effective means of water removal and thus offer a process for the simultaneous **dehydration** and **desulphurization** of gas. Gas that has an excessively high water content may require upstream dehydration, however. A portion of the natural gas may also be lost by the adsorption of hydrocarbon components by the sieve. In this process, unsaturated hydrocarbon components, such as **olefins** and **aromatics**, tend to be strongly adsorbed by the molecular sieves. The molecular sieves are susceptible to poisoning by such chemicals as glycols and require thorough gas cleaning methods before the adsorption step.

In another process, a **membrane** – based process for upgrading natural gas that contains C_3 + hydrocarbons and/or acid gas is described. The conditioned natural gas can be used as fuel for gas – powered equipment, including compressors, in the gas field or the proces sing plant. Optionally, the process can be used to produce natural gas liquids.

Process Selection

Each of the previous treating processes has advantages relative to the others for certain applications; therefore, in selection of the appropriate process, the following facts should be considered.

(1) Air pollution regulations regarding sulfur compound disposal and/or **Tail Gas Clean Up** (**TGCU**) requirements.

(2) Type and concentration of impurities in the sour gas.

(3) Specifications for the residue gas.

(4) Specifications for the acid gas.

(5) Temperature and pressure at which the sour gas is available and at which the sweet gas must be delivered.

(6) Volume of gas to be processed.

(7) Hydrocarbon composition of the gas.

(8) Selectivity required for acid gas removal.

(9) Capital cost and operating cost.

(10) Royalty cost for process.

(11) Liquid product specifications.

(12) Disposal of by – products considered hazardous chemicals.

Decisions in selecting a gas treating process can often be simplified by gas composition and operating conditions. High partial pressures (50 psia) of acid gases enhance the probability of using a physical solvent. The presence of significant quantities of heavy hydrocarbons in the feed discourages using physical solvents. Low partial pressures of acid gases and low outlet specifications generally require the use of amines for adequate treating. Process selection is not easy and a number of variables must be weighed prior to making a process selection. After preliminary assessment, a study of relevant alternatives is usually required.

In general, batch and amine processes are used for over 90% of all onshore wellhead applications. Amine processes are preferred, because of the lower operating cost, when the chemical cost for this process is prohibitive, and justifies the higher equipment cost. The key determinant is the sulfur content of the feed gas. When the sulfur content is below 20 – pound sulfur per day, batch processes are more economical, and when the sulfur content is over 100 – pound sulfur per day amine solutions are preferred.

Sulfur Recovery Processes

The side stream from acid gas treating units consists mainly of hydrogen sulfide/or carbon dioxide. Carbon dioxide is usually vented to the atmosphere but sometimes is recovered for CO_2 floods. Hydrogen sulfide could be routed to an **incinerator** or flare, which would convert the H_2S to SO_2. The release of H_2S to the atmosphere may be limited by environmental regulations. There are many specific restrictions on these limits, and the allowable limits are revised periodically. In any case, environmental regulations severely restrict the amount of H_2S that can be vented or flared in the regeneration cycle.

Most sulfur recovery processes use chemical reactions to oxidize H_2S and produce elemental sulfur. These processes are generally based either on the reaction of H_2S and O_2 or H_2S and SO_2. Both reactions yield water and elemental sulfur. These processes are licensed and involve specialized catalysts and/or solvents. These processes can be used directly on the produced gas stream. Where large flow rates are encountered, it is more common to contact the produced gas stream with a chemical or physical solvent and use a direct conversion process on the acid gas liberated in the regeneration step.

There are two common methods of sulfur recovery: **liquid redox** and **Claus sulfur recovery processes**.

Liquid redox sulfur recovery processes are **liquid – phase oxidation processes** that use a dilute aqueous solution of iron or **vanadium** to remove H_2S selectively by chemical absorption from sour gas streams. These processes can be used on relatively small or dilute H_2S stream to recover sulfur from the acid gas stream or, in some cases, can be used in place of an acid gas removal process. The mildly alkaline lean liquid scrubs the H_2S from the inlet gas stream, and the catalyst oxidizes the H_2S to elemental sulfur. The reduced catalyst is regenerated by contact with air in the oxidizer(s). Sulfur is removed from the solution by **flotation** or settling, depending on the process.

The Claus sulfur recovery process is the most widely used techno log y for recovering elemental sulfur from sour gas. The Claus process is used to recover sulfur from the amine regenerator vent gas stream in plants where large quantities of sulfur are present. However, this process is used to treat gas streams with a maximum H_2S content of 15%. The chemistry of the units involves partial oxidation of hydrogen sulfide to sulfur dioxide and the catalytically promoted reaction of H_2S and SO_2 to produce elemental sulfur.

🔲 Words and Expressions

reagent	反应物
sodium	钠
potassium	钾
arsenite	亚砷酸盐

arsenate	砷酸盐
sieve	筛
dehydration	脱水
desulphurization	脱硫
olefin	烯烃
aromatic	芳香化合物
membrane	膜
incinerator	焚烧装置;炉
vanadium	钒
flotation	悬浮

Phrases and Expressions

potassium phosphate	磷酸钾
tripotassium phosphate	磷酸钾(K_3PO_4)
absorber tower	吸收塔
reactivator tower	再生塔
alkazid process	碱处理法
potassium carbonate	碳酸钾
Giammarco – Vetrocoke	砷碱
arsenic trioxide	三氧化二砷
selenous acid	亚硒酸
tellurous acid	亚碲酸
Tail Gas Clean Up	尾气净化
liquid redox	液相氧化还原法
Claus sulfur recovery process	克劳斯回收法
liquid – phase oxidation process	液相氧化还原法

Language Focus

1. The treatment solution is a water solution of tripotassium phosphate (K_3PO_4), which is circulated through an absorber tower and a reactivator tower in much the same way as the ethanolamine is circulated in the Girbotol process; the solution is regenerated thermally.

(参考译文:处理溶液是磷酸钾(K_3PO_4)水溶液,该溶液在吸收塔和再生塔中循环,与乙醇胺法中的乙醇胺循环方式是一样的;溶液可以加热再生。)

本句中"which"引导非限制性定语从句,指代 "a water solution of tripotassium phosphate";"in much the same way as"是固定表达,意为"与……方式一样"。

🔲 Reinforced Learning

Ⅰ. **Answer the following questions for a comprehension of the text.**

1. What are other processes of acid gas treatment?

2. What facts should be considered in selection of the appropriate process?

3. What is the key process determinant?

4. What are two common methods of sulfur recovery?

5. What is the most widely used techno log y for recovering elemental sulfur from sour gas?

Ⅱ. **Multiple choice**: **choose the correct one from the alternative answers to give the exact meaning of the words.**

1. Molecular sieves are highly selective for the removal of hydrogen sulfide (as well as other sulfur compounds) from gas streams and over a <u>continuously</u> high absorption efficiency.

    A. contemporary   B. continental   C. sufficient   D. repeatedly

2. Gas that has an excessively high water content may require <u>upstream</u> dehydration, however.

    A. updated   B. upriver

    C. upgraded   D. downstream

3. The presence of significant quantities of heavy hydrocarbons in the feed <u>discourages</u> using physical solvents.

    A. encourages   B. disapproves

    C. disciplines   D. discharges

4. After <u>preliminary</u> assessment, a study of relevant alternatives is usually required.

    A. preview   B. premise   C. preceding   D. precise

5. In general, batch and amine processes are used for over 90% of all <u>onshore</u> wellhead applications.

    A. continental   B. offshore   C. shoreline   D. ocean

6. The key <u>determinant</u> is the sulfur content of the feed gas.

A. factor
B. element

C. point
D. determiner

7. There are many specific restrictions on these limits, and the allowable limits are revised <u>periodically</u>.

A. occasionally
B. optionally

C. eternally
D. continuously

8. In any case, environmental regulations <u>severely</u> restrict the amount of H_2S that can be vented or flared in the regeneration cycle.

A. definitely
B. effectively

C. strictly
D. excessively

9. Both reactions <u>yield</u> water and elemental sulfur.

A. give up　　　B. generate　　　C. exclude　　　D. encounter

10. These processes are <u>licensed</u> and involve specialized catalysts and/or solvents.

A. driven
B. operated

C. regenerated
D. authorized

Ⅲ. Multiple choice：read the four suggested translations and choose the best answer.

1. The process using <u>potassium phosphate</u> is known as phosphate desulfhurization and is used in the same way as the Girbotol process to remove acid gases from liquid hydrocarbons as well as from gas streams.

A. 砷酸盐　　　B. 碳酸钠　　　C. 碳酸钾　　　D. 磷酸钾

2. Other processes include the <u>Alkazid process</u>.

A. 热碳酸钾法
B. 碱处理法

C. 砷碱法
D. 乙醇胺法

3. The <u>Giammarco – Vetrocoke process</u> is used for hydrogen sulfide and/or carbon dioxide removal.

A. 砷碱法
B. 磷酸钾脱硫

C. 碱处理法
D. 热碳酸钾法

4. In another process, a <u>membrane – based process</u> for upgrading natural gas that contains C3 + hydrocarbons and/or acid gas is described.

A. 砷碱法
B. 胺处理

C. 分子筛法
D. 膜分离技术

5. The conditioned natural gas can be used as fuel for gas – powered e-quipment, including compressors, in the gas field or the proces sing plant.

    A. 气驱动仪器   B. 燃气动力设备

    C. 气田处理厂   D. 净化气体机器

6. Most sulfur recovery processes use chemical reactions to oxidize H_2S and produce elemental sulfur.

    A. 单元硫   B. 单质硫   C. 因素硫   D. 硫成分

7. Where large flow rates are encountered, it is more common to contact the produced gas stream with a chemical or physical solvent and use a direct conversion process on the acid gas liberated in the regeneration step.

    A. 直接转换工艺   B. 间接转接工艺

    C. 直接转化过程   D. 直接反转流程

8. The mildly alkaline lean liquid scrubs the H_2S from the inlet gas stream, and the catalyst oxidizes the H_2S to elemental sulfur.

    A. 液相氧化   B. 稀溶液

    C. 弱碱性液体   D. 溶液悬浮

9. The reduced catalyst is regenerated by contact with air in the oxidizer(s).

    A. 酸性气体   B. 回收胺

    C. 减少的催化剂   D. 被还原的催化剂

10. Sulfur is removed from the solution by flotation or settling, depending on the process.

    A. 向上或向下   B. 漂浮或下沉

    C. 悬浮或沉降   D. 上升或下降

Ⅳ. Put the following sentences into Chinese.

1. The hot potassium carbonate process decreases the acid content of natural and refinery gas from as much as 50% to as low as 0. 5% and operates in a unit similar to that used for amine treating.

2. In the hydrogen sulfide removal sec tion, the reagent consists of sodium or potassium carbonates containing a mixture of arsenites and arsenates; the carbon dioxide removal sec tion utilizes hot aqueous alkali carbonate solution activated by arsenic trioxide or selenous acid or tellurous acid.

3. Where large flow rates are encountered, it is more common to contact the produced gas stream with a chemical or physical solvent and use a direct conversion process on the acid gas liberated in the regeneration step.

4. When the sulfur content is below 20 – pound sulfur per day, batch processes are more economical, and when the sulfur content is over 100 – pound sulfur per day amine solutions are preferred.

5. Liquid redox sulfur recovery processes are liquid – phase oxidation processes that use a dilute aqueous solution of iron or vanadium to remove H_2S selectively by chemical absorption from sour gas streams.

V. Put the following paragraphs into Chinese.

1. Molecular sieves are highly selective for the removal of hydrogen sulfide from gas streams and over a continuously high absorption efficiency. They are also an effective means of water removal and thus offer a process for the simultaneous dehydration and desulphurization of gas. Gas that has an excessively high water content may require upstream dehydration, however. A portion of the natural gas may also be lost by the adsorption of hydrocarbon components by the sieve. In this process, unsaturated hydrocarbon components, such as olefins and aromatics, tend to be strongly adsorbed by the molecular sieves. The molecular sieves are susceptible to poisoning by such chemicals as glycols and require thorough gas cleaning methods before the adsorption step.

2. Decisions in selecting a gas treating process can often be simplified by gas composition and operating conditions. High partial pressures of acid gases enhance the probability of using a physical solvent. The presence of significant quantities of heavy hydrocarbons in the feed discourages using physical solvents. Low partial pressures of acid gases and low outlet specifications generally require the use of amines for adequate treating. Process selection is not easy and a number of variables must be weighed prior to making a process selection. After preliminary assessment, a study of relevant alternatives is usually required.

Chapter 6   Natural Gas Transportation

6.1   Raw Gas Transmission

🔲 Guidance to Reading

*For offshore gas fields, where space is limited, natural gas proces sing is often kept to a minimum, and the total production has to be transported via multiphase pipelines which may contain a three – phase mixture of hydrocarbon **condensate**, water, and natural gas. To transport multiphase well fluid, single pipelines can be used to substitute for separate pipelines to reduce **capital expenditure**. Therefore, **optimal** design of such pipelines is necessary, and the designer must secure a safe, efficient, and reliable transmission system throughout the design life.*

🔲 Text

Natural gas is often found in places where there is no local market, such as in the many offshore fields around the world. For natural gas to be available to the market, it must be gathered, processed, and transported. Quite often, collected natural gas (**raw gas**) must be transported over a substantial distance in pipelines of different sizes. These pipelines vary in length between hundreds of feet to hundreds of miles, across **undulating terrain** with varying temperature conditions. Liquid condensation in pipelines commonly occurs because of the multicomponent nature of the transmitted natural gas and its associated phase behavior to the inevitable temperature and pressure changes that occur along the pipeline. Condensation subjects the raw gas transmission pipeline to two – phase, gas/condensate, flow transport. The problem of optimal design of such pipelines becomes **accentuated** for offshore gas fields, where space is limited and proces sing often is kept to a minimum; therefore, total production has to be transported via multiphase pipelines. These lines lie at the bottom of the ocean in horizontal and near – horizontal positions and may contain a three – phase mixture of hydrocarbon condensate, water (occurring naturally in the reservoir), and natural gas flowing through them.

Multiphase Transportation

Multiphase transportation technology has become increasingly important for developing marginal fields, where the trend is to economically transport unprocessed well fluids via existing infrastructures, maximizing the rate of return and minimizing both capital expenditure (CAPEX) and **operational expenditure** (OPEX). In fact, by transporting multiphase well fluid in a single pipeline, separate pipelines and receiving facilities for separate phases, costing both money and space, are eliminated, which reduces capital expenditure. However, phase separation and reinjection of water and gas save both capital expenditure and operating expenditure by reducing the size of the fluid transport/handling facilities and the maintenance required for the pipeline operation.

In **thermal – hydraulic design** of multiphase flow transmission systems, the system designer is faced with several challenges associated with multiphase flow, which can significantly change design requirements. The goal of any pipeline designer is to sec ure "flow assurance," i. e. , the transmission system must operate in a safe, efficient, and reliable manner throughout the design life. Failure to do so has significant economic consequences, particularly for offshore gas production systems. "Flow assurance" covers the whole range of possible flow problems in pipelines, including both multiphase flow and fluid – related effects such as gas hydrate formation, wax and **asphaltene deposition** on walls, **corrosion**, **erosion**, **scaling**, emulsions, **foaming**, and severe **slugging**.

Gas Hydrates

A gas hydrate is an ice – like crystalline solid called a **clathrate**, which occurs when water molecules form a cage – like structure around smaller guest molecules. The most common guest molecules are methane, ethane, propane, **isobutane**, **normal butane**, nitrogen, carbon dioxide, and **hydrogen sulfide**, of which methane occurs most abundantly in natural hydrates. Several different hydrate structures are known. The two most common are structure I and structure II. Type I forms with smaller gas molecules such as methane, ethane, hydrogen sulfide, and carbon dioxide, whereas structure II is a **diamond lattice**, formed by large molecules such as propane and isobutane.

However, nitrogen, a relatively small molecule, also forms a type II hydrate. In addition, in the presence of free water, the temperature and pressure can also govern the type of hydrate structure, where the hydrate structure may change from structure II hydrate at low temperatures and pressures to structure I hydrate at high pressures and temperatures. It should be noted that **n – butane** does form a hydrate, but is very unstable. However, it will form a stabilized hydrate in the presence of small "help" gases such as methane or nitrogen. It has been assumed that normal paraffin molecules larger than n – butane are non-hydrate formers.

While many factors influence hydrate formation, the two major conditions that promote hydrate formations are the gas being (1) at the appropriate temperature and pressure and (2) at or below its water dew point. Note that free water is not necessary for hydrate formation, but it certainly enhances hydrate formation. Other factors that affect, but are not necessary for, hydrate formation include **turbulence**, **nucleation sites**, **surface for crystal formation**, **agglomeration**, and the **salinity** of the system. The exact temperature and pressure at which hydrates form depend on the composition of the gas and the water. For any particular composition of gas at a given pressure, there is a temperature below which hydrates will form and above which hydrates will not form. As the pressure increases, the hydrate formation temperature also increases. As a general rule, when the pressure of the gas stream increases or as the gas becomes colder, the tendency to form hydrates increases. Hence, many gas – handling systems are at significant risk of forming hydrate plugs during shut – in and subsequent start – up.

Although gas hydrates may be of potential benefit both as an important source of hydrocarbon energy and as a means of storing and transmitting natural gas, they represent a severe operational problem, as the hydrate crystals may deposit on the pipe wall and accumulate as large plugs that can completely block pipelines, shutting in production. Acceleration of these plugs due to a pressure gradient can also cause considerable damage to production facilities. In addition, the remediation of hydrate blockages can present significant technical difficulties with major cost implications. Because of these problems, methods of preventing hydrate solids development in gas production systems have been of considerable interest for a number of years.

The multiphase fluid produced at the wellhead will normally be at a high pressure and a moderate temperature. As the fluid flows through the pipelines, it becomes colder, which means such pipelines could experience hydrates at some point in their operating envelope. For this reason, the hydrate formation in gas transmission pipelines should be prevented effectively and economically to guarantee that the pipelines operate normally. Control of hydrates relies on keeping the system conditions out of the region in which hydrates are stable. It may be possible to keep the fluid warmer than the hydrate formation temperature (with the inclusion of a suitable margin for safety) or operate at a pressure less than the hydrate formation pressure.

Words and Expressions

condensate	凝析油
optimal	最佳的
accentuate	加重
corrosion	腐蚀
erosion	冲蚀
scaling	结垢
foaming	起泡
slugging	堵塞
clathrate	冰状晶体
isobutane	异丁烷
n – butane	正丁烷
turbulence	紊流
agglomeration	聚结
salinity	矿化度

Phrases and Expressions

capital expenditure	固定资产投资
raw gas	湿气
undulating terrain	高低起伏的地势
operational expenditure	操作成本
thermal – hydraulic design	热力和水力设计
asphaltene deposition	沥青沉积
normal butane	正丁烷

hydrogen sulfide	硫化氢
diamond lattice	金刚石型晶格
nucleation site	结晶位置
surface for crystal formation	结晶表面

Language Focus

1. Liquid condensation in pipelines commonly occurs because of the multicomponent nature of the transmitted natural gas and its associated phase behavior to the inevitable temperature and pressure changes that occur along the pipeline.

（参考译文：天然气组分繁多，在管道运输中会因温度和压力的变化而发生相态变化，在管道中出现凝析液。）

本句中"because of"引导原因状语"the multicomponent nature of the transmitted natural gas and its associated phase behavior to the inevitable temperature and pressure changes that occur along the pipeline"，其中"that"引导限制性定语从句"occur along the pipeline"，先行词是"changes"，指代"the inevitable temperature and pressure changes"。

2. The problem of optimal design of such pipelines becomes accentuated for offshore gas fields, where space is limited and proces sing often is kept to a minimum; therefore, total production has to be transported via multiphase pipelines.

（参考译文：海上气田空间受限，所有产出物经过简单处理后即输入管道，在管道内呈现多相流动状态，因此，对于此类管道，优化设计显得尤为重要。）

本句中"where"引导非限制性定语从句，指代"in offshore gas fields"；"therefore"是副词，意为"因此，所以"。

3. Multiphase transportation technology has become increasingly important for developing marginal fields, where the trend is to economically transport unprocessed well fluids via existing infrastructures, maximizing the rate of return and minimizing both capital expenditure (CAPEX) and operational expenditure (OPEX).

（参考译文：混相输送技术对于海上边际油气田开发越来越重要，海上油气开发要利用现有的基础设施经济、有效地输送未处理的井筒流体，实现收益率最大化和固定资产投资及操作成本最小化。）

本句中"where"引导非限制性定语从句"the trend is to economically transport unprocessed well fluids via existing infrastructures, maximizing the rate of return and minimizing both capital expenditure（CAPEX）and operational expenditure（OPEX）"，"where"相当于"in developing marginal fields"；"via"作介词，意为"通过"；后面的部分"maximizing the rate of return and minimizing both capital expenditure（CAPEX）and operational expenditure（OPEX）"为伴随状语，表示结果。

4. In addition, in the presence of free water, the temperature and pressure can also govern the type of hydrate structure, where the hydrate structure may change from structure II hydrate at low temperatures and pressures to structure I hydrate at high pressures and temperatures.

（参考译文：此外，由于自由水的存在，温度和压力也会对水合物的结构产生影响，低温低压下的Ⅱ型水合物会随着温度和压力的升高逐渐转换成Ⅰ型。）

本句中"In addition"表示"另外，此外"；"where"引导非限制性定语从句"the hydrate structure may change from structure II hydrate at low temperatures and pressures to structure I hydrate at high pressures and temperatures"，"where"相当于"in the presence of free water"；"change from...to..."为固定搭配。

5. Although gas hydrates may be of potential benefit both as an important source of hydrocarbon energy and as a means of storing and transmitting natural gas, they represent a severe operational problem, as the hydrate crystals may deposit on the pipe wall and accumulate as large plugs that can completely block pipelines, shutting in production.

（参考译文：天然气水合物无论是作为一种重要的碳氢化合物能源的来源，还是作为一种集输天然气的新方法，都具有可观的前景，但是水合物的形成却给生产造成了严重问题：水合物晶体可能在管壁沉降、逐渐累积甚至堵塞整个管道，最终导致生产中断。）

本句中"Although"引导让步状语从句"gas hydrates may be of potential benefit both as an important source of hydrocarbon energy and as a means of storing and transmitting natural gas"，其中两个"as"为介词，意思是"作为"，"both...and..."为固定搭配；主句是"they represent a severe operational problem"；第三个"as"作连词，引导原因状语从句"as the hydrate crystals may..."；第四个"as"为介词，"accumulate as large plugs"意为"累积成大栓

塞","that"指代"large plugs",引导限制性定语从句"can completely block pipelines";"shutting in production"为伴随状语,表示结果。

🔛 Reinforced Learning

I. Answer the following questions for a comprehension of the text.

1. Why has multiphase transportation technology become more important for developing marginal fields?

2. What is the goal of any pipeline designer?

3. What is a gas hydrate?

4. What structure does a hydrate have?

5. How to prevent the hydrate formation in gas transmission pipelines effectively and economically?

II. Multiple choice: choose the correct one from the alternative answers to give the exact meaning of the words.

1. For natural gas to be <u>available</u> to the market, it must be gathered, processed, and transported.

    A. popular   B. sold   C. interesting   D. appealing

2. The problem of <u>optimal</u> design of such pipelines becomes accentuated for offshore gas fields, where space is limited and proces sing often is kept to a minimum; therefore, total production has to be transported via multiphase pipelines.

    A. best   B. optional   C. chosen   D. worse

3. Multiphase transportation techno log y has become increa singly important for developing <u>marginal</u> fields, where the trend is to economically transport unprocessed well fluids via existing infrastructures, maximizing the rate of return and minimizing both capital expenditure (CAPEX) and operational expenditure (OPEX).

    A. oil   B. gas   C. inland   D. offshore

4. In thermal – hydraulic design of multiphase flow transmission systems, the system designer is faced with several <u>challenges</u> associated with multiphase flow, which can significantly change design requirements.

    A. problems   B. designs   C. methods   D. schemes

5. <u>In addition</u>, in the presence of free water, the temperature and pressure

can also govern the type of hydrate structure, where the hydrate structure may change from structure II hydrate at low temperatures and pressures to structure I hydrate at high pressures and temperatures.

    A. However   B. Besides

    C. Particularly   D. Nevertheless

6. It has been <u>assumed</u> that normal paraffin molecules larger than n - butane are nonhydrate formers.

    A. resumed   B. assured

    C. presumed   D. believed

7. Note that free water is not necessary for hydrate formation, but it certainly <u>enhances</u> hydrate formation.

    A. promotes   B. improves

    C. results   D. increases

8. The exact temperature and pressure at which hydrates form depend on the <u>composition</u> of the gas and the water.

    A. constitution   B. mixing

    C. integration   D. separation

9. <u>Hence</u>, many gas - handling systems are at significant risk of forming hydrate plugs during shut - in and subsequent start - up.

    A. Although   B. Besides   C. However   D. Therefore

10. Acceleration of these plugs <u>due to</u> a pressure gradient can also cause considerable damage to production facilities.

    A. result in   B. in order to

    C. owing to   D. in case of

Ⅲ. Multiple choice: read the four suggested translations and choose the best answer.

1. These pipelines vary in length between hundreds of feet to hundreds of miles, across <u>undulating terrain</u> with varying temperature conditions.

    A. 土地肥沃   B. 地广人稀

    C. 景色宜人   D. 地势高低起伏

2. Multiphase transportation technology has become increasingly important for developing marginal fields, where the trend is to economically transport unprocessed well fluids via existing infrastructures, maximizing the rate of return

and minimizing both capital expenditure (CAPEX) and operational expenditure (OPEX).

　　A. 开支预算和操作预算　　　B. 资本运营和管理经费

　　C. 固定资产投资和操作成本　D. 资本操作和操作花费

　　3. In fact, by transporting multiphase well fluid in a single pipeline, separate pipelines and receiving facilities for separate phases, costing both money and space, are eliminated, which reduces capital expenditure.

　　A. 多层　　　　B. 混相　　　C. 混合　　　　D. 多级

　　4. The goal of any pipeline designer is to sec ure "flow assurance," i. e., the transmission system must operate in a safe, efficient, and reliable manner throughout the design life.

　　A. 流动保险　　　　　　　　B. 流动安全

　　C. 确保流动　　　　　　　　D. 肯定流动

　　5. "Flow assurance" covers the whole range of possible flow problems in pipelines, including both multiphase flow and fluid – related effects such as gas hydrate formation, wax and asphaltene deposition on walls, corrosion, erosion, scaling, emulsions, foaming, and severe slugging.

　　A. 管壁结蜡　　　　　　　　B. 腐蚀冲蚀

　　C. 沥青沉积　　　　　　　　D. 乳化起泡

　　6. A gas hydrate is an ice – like crystalline solid called a clathrate, which occurs when water molecules form a cage – like structure around smaller guest molecules.

　　A. 天然气水合物　　　　　　B. 甲烷水合物

　　C. 金刚石晶格　　　　　　　D. 冰状晶体

　　7. The most common guest molecules are methane, ethane, propane, isobutane, normal butane, nitrogen, carbon dioxide, and hydrogen sulfide, of which methane occurs most abundantly in natural hydrates.

　　A. 二氧化碳　　B. 硫化氢　　C. 异丁烷　　　　D. 正丁烷

　　8. It should be noted that n – butane does form a hydrate, but is very unstable.

　　A. 乙烷　　　　B. 丙烷　　　C. 异丁烷　　　　D. 正丁烷

　　9. Other factors that affect, but are not necessary for, hydrate formation include turbulence, nucleation sites, surface for crystal formation, agglomeration, and the salinity of the system.

A. 结晶位置   B. 晶体聚结

C. 结晶表面   D. 系统矿化度

10. Hence, many gas-handling systems are at significant risk of forming hydrate plugs during <u>shut-in</u> and subsequent <u>start-up</u>.

A. 启动 关闭   B. 关掉 打开

C. 关上 启动   D. 关闭 重启

IV. Put the following sentences into Chinese.

1. Multiphase transportation techno log y has become increa singly important for developing marginal fields, where the trend is to economically transport unprocessed well fluids via existing infrastructures, maximizing the rate of return and minimizing both capital expenditure (CAPEX) and operational expenditure (OPEX).

2. The goal of any pipeline designer is to sec ure "flow assurance," i. e. , the transmission system must operate in a safe, efficient, and reliable manner throughout the design life.

3. "Flow assurance" covers the whole range of possible flow problems in pipelines, including both multiphase flow and fluid-related effects such as gas hydrate formation, wax and asphaltene deposition on walls, corrosion, erosion, scaling, emulsions, foaming, and severe slugging.

4. Type I forms with smaller gas molecules such as methane, ethane, hydrogen sulfide, and carbon dioxide, whereas structure II is a diamond lattice, formed by large molecules such as propane and isobutane.

5. Although gas hydrates may be of potential benefit both as an important source of hydrocarbon energy and as a means of storing and transmitting natural gas, they represent a severe operational problem, as the hydrate crystals may deposit on the pipe wall and accumulate as large plugs that can completely block pipelines, shutting in production.

V. Put the following paragraphs into Chinese.

1. Natural gas is often found in places where there is no local market, such as in the many offshore fields around the world. For natural gas to be available to the market, it must be gathered, processed, and transported. Quite often, raw gas must be transported over a substantial distan ce in pipelines of different sizes. These pipelines vary in length between hundreds of feet to

hundreds of miles, across undulating terrain with varying temperature conditions. Liquid condensation in pipelines commonly occurs because of the multi-component nature of the transmitted natural gas and its associated phase behavior to the inevitable temperature and pressure changes that occur along the pipeline.

2. The exact temperature and pressure at which hydrates form depend on the composition of the gas and the water. For any particular composition of gas at a given pressure, there is a temperature below which hydrates will form and above which hydrates will not form. As the pressure increases, the hydrate formation temperature also increases. As a general rule, when the pressure of the gas stream increases or as the gas becomes colder, the tendency to form hydrates increases. Hence, many gas – handling systems are at significant risk of forming hydrate plugs during shut – in and subsequent start – up.

6.2　Hydrate Prevention Techniques

［巾］ Guidance to Reading

Most natural gas contains substantial amounts of water vapor at the time it is produced from a well or separated from an associated crude – oil stream. Water vapor must be removed from the gas stream because it will condense into liquid and may cause hydrate formation as the gas is cooled from the high reservoir temperature to the cooler surface temperature. Liquid water almost always accelerates corrosion, and the solid hydrates may pack solidly in gas – gathering systems, resulting in partial or complete blocking of flow lines. As a result, it has brought home to us the importance of hydrate prevention method study, and proper selection of hydrate prevention methods for subsea gas transmission pipelines is more challenging.

［巾］ Text

Although there are several methods to avoid hydrate formation, depending on the possible location of a gas hydrate plug, some of the techniques to **remediate** onshore hydrates (e. g. , installation of line heaters and line **depressurization**) may not be practical in long and high pressure, subsea gas transmission pipelines. There are few methods of preventing hydrate formation in offshore transmission systems. The permanent solution is removal of water prior to pipeline transportation, using a large offshore **dehydration** plant that is not often the most cost – effective solution. In general, two methods are applicable at the well site, namely thermal and chemical.

Thermal Methods

Thermal methods use either conservation or introduction of heat in order to maintain the flowing mixture outside the hydrate formation range. Heat conservation is common practice and is accomplished through **insulation**. This method can be **feasible** for some subsea applications depending on the fluid being transported, the tieback distance, and topside capabilities of the host platform. The design of such conservation systems typically seeks a balance among the high cost of the insulation, the intended operability of the system, and the acceptable risk level.

A number of different concepts are available for introducing additional heat to a pipeline. The simplest is an **external hot – water jacket**, either for a **pipe – in – pipe system** or for a **bundle**. Other methods use either **conductive** or **inductive heat tracing**. There is concern over the reliability of conductive systems. An **electrical resistance heating system** may be desirable for long offset systems, where available insulation is **insufficient**, or for shut – in conditions. The ability to heat during production depends on the specific electrical heating **implementation**.

Such systems provide environmentally friendly fluid temperature control without flaring for pipeline depressurization. The effect is also an increase in production as there is no time lost by unnecessary depressurization, **pigging**, heating – medium circulation, or removal of hydrate blockage. However, it is still difficult to persuade operators to install an acting heating system.

Chemical Inhibition

An alternative to the thermal processes is chemical inhibition. **Chemical inhibitors** are injected at the wellhead and prevent hydrate formation by depressing the hydrate temperature below that of the pipeline operating temperature. Chemical injection systems for subsea lines have a rather high capital expenditure price tag associated with them, in addition to the often high operating cost of chemical treatment. However, hydrate inhibition using chemical inhibitors is still the most widely used method for unprocessed gas streams, and the development of alternative, cost – effective, and environmentally acceptable hydrate inhibitors is a techno log ical challenge for the gas production industry.

Types of Inhibitors

Traditionally, the most common **chemical additives** used to control hydrates in gas production systems have been methanol, **ethylene glycol**, or **triethylene glycol** at a high enough concentration. Increasing salt content in the produced brine (by injecting **electrolyte solutions** such as **sodium chloride**, **calcium chloride**, and **potassium chloride**) can also provide some hydrate temperature suppression, but normally this alone is not sufficient to avoid hydrates in the subsea gas production systems. In some cases, **blended inhibitors** of methanol/glycols and electrolyte(s) are preferred for subsea applications.

The inhibitor selection process often involves comparison of many factors, including capital/operating cost, physical properties, safety, **corrosion inhibition**, and gas dehydration capacity. However, a primary factor in the selection process is whether the spent chemical will be recovered, regenerated, and reinjected. However, losses to the vapor phase can be **prohibitive**, in which case operators select **monoethylene glycol**. Often when applying this inhibitor, there is a significant expense associated with the cost of "lost" methanol. However, because methanol has lower **viscosity** and lower surface tension it makes for an effective separation from the gas phase at **cryogenic** conditions (below – 13 ℉) and is usually preferred. Because glycols are expensive inhibitors, there is a definite need for extra, costly, and space – consuming onshore or offshore plants for their regeneration. Therefore, it would be useful to develop new hydrate **depressants**, which can be used at much lower concentrations and therefore much lower cost.

Two new types of low – dosage inhibitors have been developed, that will enable the subsea gas transmission pipelines to handle increased gas volumes without additional glycol injection or extra glycol recovery units. These new hydrate inhibitors can lead to very substantial cost savings, not only for the reduced cost of the new inhibitor, but also in the size of the injection, pumping, and storage facilities. These new hydrate inhibitors, called "**low – dosage hydrate inhibitors**" (**LDHIs**), form the basis of a technique that does not operate by changing the thermodynamic conditions of the system.

In contrast to other types of inhibitors, AAs, which are surface – active chemicals, do not prevent the formation of hydrate crystals but keep the particles small and well dispersed so that fluid vis cos ity remains low, allowing the hydrates to be transported along with the produced fluids. AAs performance is relatively independent of time. In addition, AAs appear to be effective at more extreme conditions than KHIs, which makes these products of interest to operators looking for cost – effective hydrate control in deepwater fields. These additives are currently applied in the Gulf of Mexico, the North Sea, and west Africa. However, they have mainly limitations in terms of water cut, where they require a continuous oil phase and therefore are only applicable at lower water cuts. The maximum water cut is expected to be between 40 and 50%. This limitation is caused by the **rheological properties** of suspensions with high solid fraction and may depend on **flow regime** conditions.

As stated earlier, the choice between inhibitor alternatives must be based on physical limitations as well as economics. However, operating conditions may also limit the number of available choices. For example, in a project carried out by Baker Petrolite it was shown that under severe conditions, the required dosage of an **antiagglomerator** unlike thermodynamic and **kinetic inhibitors** does not increase as the degree of subcooling increases. Therefore, this method of treatment would be a cost – effective solution for the control of gas hydrates.

Prediction of Inhibitor Requirements

The inhibitor must be present in a minimum concentration to avoid hydrate formation. Accurate prediction of this minimum inhibitor concentration is required for costeffective design and operation of multiphase pipelines.

Design of Injection Systems

Proper design of an inhibitor injection system is a complex task that involves optimum inhibitor selection, determination of the necessary injection rates, pump sizing, and pipeline diameters. Inhibitors for a subsea gas transmission system are selected before gas production is started on the facility. This makes inhibitor selection difficult, as a large number of factors, including brine composition, temperature, and pressure, that affect the performance of inhibitors are unknown. Therefore, at this stage, an appropriate multiphase flow simulation package must be used to calculate some of the unknown necessary variables, which are required for injection systems design.

🔲 **Words and Expressions**

remediate	修复
depressurization	降压
dehydration	脱水
insulation	隔热
feasible	可行的
bundle	丛式管
conductive	传导式
inductive	感应式
insufficient	不足的
implementation	实施
pigging	清蜡
prohibitive	禁止的
vis cos ity	黏度
cryogenic	低温的
depressant	抑制剂
antiagglomerator	反凝聚剂

🔲 **Phrases and Expressions**

external hot – water jacket	外加水套炉
pipe – in – pipe system	双层套管
heat tracing	伴热
electrical resistance heating system	电阻加热系统
chemical inhibitor	化学抑制剂
chemical additive	化学助剂
ethylene glycol	乙二醇
triethylene glycol	三甘醇
electrolyte solution	电解质溶液
sodium chloride	氯化钠
calcium chloride	氯化钙
potassium chloride	氯化钾
blended inhibitor	混合抑制剂
corrosion inhibition	防腐

monoethylene glycol	乙二醇
low – dosage hydrate inhibitors	低剂量水合物抑制剂
rheo log ical properties	流变学性质
flow regime	流动型态
kinetic inhibitor	活性抑制剂

Language Focus

1. Although there are several methods to avoid hydrate formation, depending on the possible location of a gas hydrate plug, some of the techniques to remediate onshore hydrates (e. g. , installation of line heaters and line depressurization) may not be practical in long and high pressure, subsea gas transmission pipelines.

(参考译文:由于天然气水合物形成的堵塞在集输管线中所处的位置不同,水合物的防治方法也有很多种,如安装加热炉或者降压输送,但是这些方法对于海底长距离高压集输管线可能就不适用了。)

本句中"Although"引导让步状语从句"there are several methods to avoid hydrate formation";"depending on the possible location of a gas hydrate plug"为分词短语作原因状语。

2. The permanent solution is removal of water prior to pipeline transportation, using a large offshore dehydration plant that is not often the most cost – effective solution.

(参考译文:一直以来,海上油田的防治措施是在天然气进入管线之前先用大型海上脱水装置对天然气进行脱水处理,但这种方法成本较高。)

本句中"prior to"为固定搭配,意为"在……之前";"using a large offshore dehydration plant that is not often the most cost – effective solution"为伴随状语;"that"引导限制性定语从句,指代前面的"a large offshore dehydration plant"。

3. In contrast to other types of inhibitors, AAs, which are surface – active chemicals, do not prevent the formation of hydrate crystals but keep the particles small and well dispersed so that fluid vis cos ity remains low, allowing the hydrates to be transported along with the produced fluids.

(参考译文:AAs 和其他类型的化学抑制剂不同,是一种表面活性剂,虽然不能抑制水合物晶体的形成,但是可以分散这些晶体小颗粒并防止其长大。这样,流体的黏度就会保持在较低的状态,形成的水合物就会随采出流体带走。)

本句中"in contrast to"是固定搭配,意为"与……形成鲜明对照";"which"引导非限制性定语从句,指代"AAs";"so that"是固定搭配,意为"以便,所以";"allowing the hydrates to be transported along with the produced fluids"为伴随状语。

Reinforced Learning

Ⅰ. **Answer the following questions for a comprehension of the text.**

1. What are the two applicable methods of preventing hydrate formation at the well site?

2. How to conserve heat?

3. How to select inhibitor?

4. What was shown in a project carried out by Baker Petrolite?

5. How to design an inhibitor injection system?

Ⅱ. **Multiple choice: choose the correct one from the alternative answers to give the exact meaning of the words.**

1. Although there are several methods to avoid hydrate formation, depending on the possible location of a gas hydrate plug, some of the techniques to remediate onshore hydrates may not be practical in long and high pressure, subsea gas transmission pipelines.

    A. prevent   B. cure   C. solve   D. improve

2. This method can be feasible for some subsea applications depending on the fluid being transported, the tieback distance, and topside capabilities of the host platform.

    A. impossible   B. practical   C. necessary   D. important

3. There is concern over the reliability of conductive systems.

    A. possibility   B. dependability

    C. feasibility   D. capability

4. An electrical resistance heating system may be desirable for long offset systems, where available insulation is insufficient, or for shut – in conditions.

    A. ineffective   B. inevitable

    C. inappropriate   D. inadequate

5. Chemical inhibitors are injected at the wellhead and prevent hydrate formation by <u>depressing</u> the hydrate temperature below that of the pipeline operating temperature.

    A. decreasing   B. raising

    C. improving   D. changing

6. However, losses to the vapor phase can be <u>prohibitive</u>, in which case operators select monoethylene glycol.

    A. creative   B. positive

    C. unallowable   D. unpredictable

7. However, because methanol has lower vis cos ity and lower surface tension it makes for an effective separation from the gas phase at <u>cryogenic</u> conditions (below $-13\,^{\circ}\text{F}$) and is usually preferred.

    A. hot   B. warm   C. cold   D. cool

8. <u>In contrast to</u> other types of inhibitors, AAs, which are surface – active chemicals, do not prevent the formation of hydrate crystals but keep the particles small and well dispersed so that fluid viscosity remains low, allowing the hydrates to be transported along with the produced fluids.

    A. In comparison with   B. In terms of

    C. According to   D. Different from

9. Therefore, this method of treatment would be a <u>cost – effective</u> solution for the control of gas hydrates.

    A. cost – efficient   B. high – cost

    C. low – cost   D. cost – negotiable

10. Therefore, at this stage, an appropriate multiphase flow <u>simulation</u> package must be used to calculate some of the unknown necessary variables, which are required for injection systems design.

    A. activation   B. imitation

    C. genuine   D. reality

Ⅲ. Multiple choice: read the four suggested translations and choose the best answer.

1. The simplest is an <u>external hot – water jacket</u>, either for a pipe – in – pipe system or for a bundle.

    A. 保暖外套   B. 加热外衣

    C. 外加水套炉   D. 保暖水套

2. Other methods use either conductive or inductive heat tracing.

A. 归纳或演绎   B. 导电或绝缘

C. 传导或诱导   D. 传导式或感应式

3. The effect is also an increase in production as there is no time lost by unnecessary depressurization, pigging, heating – medium circulation, or removal of hydrate blockage.

A. 导热循环   B. 循环导热

C. 循环传热介质   D. 循环导热传播

4. However, hydrate inhibition using chemical inhibitors is still the most widely used method for unprocessed gas streams, and the development of alternative, cost – effective, and environmentally acceptable hydrate inhibitors is a techno log ical challenge for the gas production industry.

A. 交替的、花钱的、环保的

B. 可替代的、低成本的、环保的

C. 可选择的、成本有效的、环境友好的

D. 可替代的、成本可控的、环境容许的

5. Traditionally, the most common chemical additives used to control hydrates in gas production systems have been methanol, ethylene glycol, or triethylene glycol at a high enough concentration.

A. 助剂   B. 燃料

C. 物质   D. 工艺

6. In some cases, blended inhibitors of methanol/glycols and electrolyte (s) are preferred for subsea applications.

A. 联合禁止   B. 融合抑制剂

C. 混合抑制剂   D. 混合抑制因子

7. As stated earlier, the choice between inhibitor alternatives must be based on physical limitations as well as economics.

A. 物理因素   B. 身体局限

C. 设备的限制   D. 助剂的劣势

8. For example, in a project carried out by Baker Petrolite it was shown that under severe conditions, the required dosage of an antiagglomerator unlike thermodynamic and kinetic inhibitors does not increase as the degree of subcooling increases.

A. 动力水合物抑制剂和活性抑制剂

B. 氯化钠和氯化钙

C. 甲醇和乙二醇

D. 烷基苯磺酸盐和烷基苯乙酸盐

9. Proper design of an inhibitor injection system is a complex task that involves optimum inhibitor selection, determination of the necessary injection rates, <u>pump sizing, and pipeline diameters</u>.

A. 泵型和管径　　　　　　　B. 抑制剂用量和管道设计

C. 抑制剂优选和注入速率　　D. 盐水组分和温度

10. This makes inhibitor selection difficult, as a large number of factors, including <u>brine composition</u>, temperature, and pressure, that affect the performance of inhibitors are unknown.

A. 盐水温度　　　　　　　　B. 抑制剂压力

C. 抑制剂影响　　　　　　　D. 盐水组分

IV. Put the following sentences into Chinese.

1. The design of such conservation systems typically seeks a balance among the high cost of the insulation, the intended operability of the system, and the acceptable risk level.

2. Chemical injection systems for subsea lines have a rather high capital expenditure price tag associated with them, in addition to the often high operating cost of chemical treatment.

3. Increa sing salt content in the produced brine can also provide some hydrate temperature suppression, but normally this alone is not sufficient to avoid hydrates in the subsea gas production systems.

4. Under severe conditions, the required dosage of an antiagglomerator unlike thermodynamic and kinetic inhibitors does not increase as the degree of subcooling increases.

5. Therefore, at this stage, an appropriate multiphase flow simulation package must be used to calculate some of the unknown necessary variables, which are required for injection systems design.

V. Put the following paragraphs into Chinese.

1. The inhibitor selection process often involves comparison of many factors, including capital/operating cost, physical properties, safety, corrosion in-

hibition, and gas dehydration capacity. However, a primary factor in the selection process is whether the spent chemical will be recovered, regenerated, and reinjected. Because glycols are expensive inhibitors, there is a definite need for extra, costly, and space – consuming onshore or offshore plants for their regeneration. Therefore, it would be useful to develop new hydrate depressants, which can be used at much lower concentrations and therefore much lower cost.

2. Proper design of an inhibitor injection system is a complex task that involves optimum inhibitor selection, determination of the necessary injection rates, pump sizing, and pipeline diameters. Inhibitors for a subsea gas transmission system are selected before gas production is started on the facility. This makes inhibitor selection difficult, as a large number of factors, including brine composition, temperature, and pressure, that affect the performance of inhibitors are unknown.

6.3   Natural Gas Dehydration

🔲 Guidance to Reading

*It is necessary to reduce and control the water content of natural gas to ensure safe proces sing and transmission. The most common methods of dehydrating natural gas are liquid **desiccant** (glycol) dehydration, solid desiccant dehydration, and **refrigeration** (cooling the gas). Using desiccant dehydrators can yield significant economic and environmental benefits, including reduced capital cost, reduced operation and maintenance cost, and minimal **VOC** and hazardous air pollutants. The selection of proper desiccant for a given application is a complex problem. Some desirable properties of desiccants are offered in this sec tion for reference.*

🔲 Text

Introduction

Natural, associated, or **tail gas** usually contains water, in liquid and/or vapor form, at source and/or as a result of sweetening with an **aqueous solution**. Operating experience and thorough engineering have proved that it is necessary to reduce and control the water content of gas to ensure safe proces sing and transmission. The major reasons for removing the water from natural gas are as follow.

(1) Natural gas in the right conditions can combine with liquid or free water to form solid hydrates that can plug **valves fittings** or even pipelines.

(2) Water can condense in the pipeline, causing **slug flow** and possible erosion and corrosion.

(3) Water vapor increases the volume and decreases the heating value of the gas.

(4) Sales gas contracts and/or pipeline specifications often have to meet the maximum water content of 7 lb H_2O per MMscf.

Pipeline drips installed near wellheads and at strategic locations along gathering and trunk lines will eliminate most of the free water lifted from the wells in the gas stream. Multistage separators can also be deployed to ensure the reduction of free water that may be present. However, removal of the water vapor that exists in solution in natural gas requires a more complex treatment. This treatment consists of "dehydrating" the natural gas, which is accomplished by lowering the dew point temperature of the gas at which water vapor will condense from the gas.

There are several methods of dehydrating natural gas. The most common of these are liquid desiccant (glycol) dehydration, solid desiccant dehydration, and refrigeration (i. e. , cooling the gas). The first two methods utilize mass transfer of the water molecule into a liquid solvent (glycol solution) or a **crystalline structure** (dry desiccant). The third method employs cooling to condense the water molecule to the liquid phase with the subsequent injection of inhibitor to prevent hydrate formation. However, the choice of dehydration method is usually between glycol and solid desiccants.

Glycol Dehydration

Among the different gas drying processes, absorption is the most common technique, where the water vapor in the gas stream becomes absorbed in a liquid solvent stream. Glycols are the most widely used absorption liquids as they approximate the properties that meet commercial application criteria. Several glycols have been found suitable for commercial application.

The commonly available glycols and their uses are described as follows.

(1) **Monoethylene glycol** (**MEG**); high vapor equilibrium with gas so tend to lose to gas phase in contactor. Use as hydrate inhibitor where it can be

recovered from gas by separation at temperatures below 50 °F.

(2)**Diethylene glycol (DEG)**; high vapor pressure leads to high losses in contactor. Low decomposition temperature requires low reconcentrator temperature (315 to 340 °F) and thus cannot get pure enough for most applications.

(3)**Triethylene glycol (TEG)**; most common. Reconcentrate at 340 – 400 °F, for high purity. At contactor temperatures in excess of 120 °F, there is a tendency to high vapor losses. Dewpoint depressions up to 150 °F are possible with stripping gas.

(4)**Tetraethylene glycol (TREG)**; more expensive than TEG but less loss at high gas contact temperatures. Reconcentrate at 400 to 430 °F.

TEG is by far the most common liquid desiccant used in natural gas dehydration. It exhibits most of the desirable criteria of commercial suitability listed here.

(1)TEG is regenerated more easily to a concentration of 98 – 99% in an atmospheric stripper because of its high boiling point and decomposition temperature.

(2)TEG has an initial theoretical decomposition temperature of 404 °F, whereas that of diethylene glycol is only 328 °F.

(3)Vaporization losses are lower than monoethylene glycol or diethylene glycol. Therefore, the TEG can be regenerated easily to the high concentrations needed to meet pipeline water dew point specifications.

(4)Capital and operating costs are lower.

Solid Desiccant Dehydration

Solid desiccant dehydration systems work on the principle of adsorption. Adsorption involves a form of adhesion between the surface of the solid desiccant and the water vapor in the gas. The water forms an extremely thin film that is held to the desiccant surface by forces of attraction, but there is no chemical reaction.

Solid desiccant dehydrators are typically more effective than glycol dehydrators, as they can dry a gas to less than 0.1 ppmV (0.05 lb/MMcf). However, in order to reduce the size of the solid desiccant dehydrator, a glycol dehydration unit is often used for bulk water removal. The glycol unit would reduce the water content to around 60 ppmV, which would help reduce the mass of solid desiccant necessary for final drying.

Using desiccant dehydrators as alternatives to glycol dehydrators can yield significant economic and environmental benefits, including reduced capital cost, reduced operation and maintenance cost, and minimal VOC and hazardous air pollutants (BTEX).

Desiccant Capacity

The **capacity** of a desiccant for water is expressed normally in mass of water adsorbed per mass of desiccant. The dynamic moisture sorption capacity of a desiccant will depend on a number of factors, such as the relative humidity of the inlet gas, the gas flow rate, the temperature of the adsorption zone, the **mesh** size of the **granule**, and the length of service and degree of contamination of the desiccant and not the least on the desiccant itself. The moisture sorption capacity is not affected by variations in pressure, except where pressure may affect the other variables listed previously. There are three capacity terms used.

(1) *Static equilibrium capacity*: The water capacity of new, virgin desiccant as determined in an equilibrium cell with no fluid flow (corresponding to the adsorption isotherm).

(2) *Dynamic equilibrium capacity*: The water capacity of desiccant where the fluid is flowing through the desiccant at a commercial rate.

(3) *Useful capacity*: The design capacity that recognizes loss of desiccant capacity with time as determined by experience and economic consideration and the fact that all of the desiccant bed can never be fully utilized.

Desiccant Selection

A variety of solid desiccants are available in the market for specific applications. Some are good only for dehydrating the gas, whereas others are capable of performing both dehydration and removal of heavy hydrocarbon components. The selection of proper desiccant for a given application is a complex problem. For solid desiccants used in gas dehydration, the following properties are desirable.

(1) High adsorption capacity at equilibrium. This lowers the required **adsorbent** volume, allowing for the use of smaller vessels with reduced capital expenditures and reduced heat input for regeneration.

(2) High selectivity. This minimizes the undesirable removal of valuable components and reduces overall operating expenses.

(3) Easy regeneration. The relatively low regeneration temperature mini-

mizes overall energy requirements and operating expenses.

(4) Low pressure drop.

(5) Good mechanical properties (such as high **crush strength**, **low attrition**, **low dust formation**, and high stability against aging). These factors lower overall maintenance requirements by reducing the frequency of adsorbent change out and minimizing **downtime** – related losses in production.

(6) Inexpensive, noncorrosive, nontoxic, chemically inert, high **bulk density** and no significant volume changes upon adsorption and desorption of water.

🔲 Words and Expressions

desiccant	干燥剂
refrigeration	冷却法
capacity	性能
mesh	筛孔大小
granule	颗粒
adsorbent	吸收剂
downtime	停工

🔲 Phrases and Expressions

VOC	挥发性有机化合物
tail gas	尾气
aqueous solution	水溶液
valve fitting	阀门辅助配线
slug flow	段塞流
crystalline structure	结晶体
monoethylene glycol	乙二醇
diethylene glycol	二甘醇
triethylene glycol	三甘醇
tetraethylene glycol	四甘醇
useful capacity	有效吸水量
crush strength	破裂强度
low attrition	低磨损
low dust formation	低粉尘

bulk density   体积密度

🔲 Language Focus

1. Natural, associated, or tail gas usually contains water, in liquid and/or vapor form, at source and/or as a result of sweetening with an aqueous solution.

（参考译文：天然气、伴生气和尾气中通常都含有以液相和气相形式存在的水，主要来源于脱硫过程中的水溶液。）

本句"natural, associated, or tail gas"中"natural, associated, or tail"三个形容词作定语修饰"gas"，意为"天然气、伴生气和尾气"；"in...form"或"in the form of..."为固定表达，意为"以……形式"；"as a result of"作介词短语，是固定表达，意为"因为，作为结果"；"sweeten"是形容词"sweet"加后缀"en"构成的动词，意为"变甜，脱硫"。

2. This treatment consists of "dehydrating" the natural gas, which is accomplished by lowering the dew point temperature of the gas at which water vapor will condense from the gas.

（参考译文：这个处理方法是给天然气"脱水"，通过降低天然气露点，使水蒸气从天然气中凝结出来。）

本句中"this treatment"是主语，"consist of"是谓语，也是固定短语，意为"包括，由……构成"；"which"引导非限制性定语从句，指代"dehydrating" the natural gas；"at which"引导限制性定语从句，相当于"at the dew point temperature"。

3. Among the different gas drying processes, absorption is the most common technique, where the water vapor in the gas stream becomes absorbed in a liquid solvent stream.

（参考译文：在不同的气体干燥方法中，吸收法是最常用的一种方法。天然气中的水蒸气可以用液体溶剂来吸收。）

本句中"among"作介词，意为"在……之中"；"where"引导非限制性定语从句，相当于"in absorption"。

🔲 Reinforced Learning

Ⅰ. Answer the following questions for a comprehension of the text.

1. What are the commonly used glycols?
2. Why is TEG most commonly used?

3. What benefits can desiccant dehydrators bring as alternatives to glycol dehydrators?

4. What are factors for dynamic moisture sorption capacity of a desiccant?

5. What properties are required for solid desiccants used in gas dehydration?

II. Multiple choice: choose the correct one from the alternative answers to give the exact meaning of the words.

1. Natural, associated, or tail gas usually contains water, in liquid and/or vapor form, at source and/or as a result of sweetening with an aqueous solution.

    A. as a way of   B. in the form of

    C. giving rise to   D. because of

2. Natural gas in the right conditions can combine with liquid or free water to form solid hydrates that can plug valves fittings or even pipelines.

    A. tools   B. suiting   C. accessories   D. pipelines

3. Sales gas contracts and/or pipeline specifications often have to meet the maximum water content of 7 lb H_2O per MMscf.

    A. agreements   B. compression

    C. contacts   D. specifications

4. Glycols are the most widely used absorption liquids as they approximate the properties that meet commercial application criteria.

    A. possess   B. approach   C. meet   D. require

5. At contactor temperatures in excess of 120 °F there is a tendency to high vapor losses.

    A. excessive   B. succeeding

    C. exceeding   D. below

6. Solid desiccant dehydration systems work on the principle of adsorption.

    A. based on the standard of   B. on the regulation of

    C. in application to   D. on perspective of

7. However, in order to reduce the size of the solid desiccant dehydrator, a glycol dehydration unit is often used for bulk water removal.

    A. glycol   B. dehydration

    C. moisture   D. big volume

8. Some are good only for dehydrating the gas, whereas others <u>are capable of</u> performing both dehydration and removal of heavy hydrocarbon components.

A. are fond of 　　　　　　　　B. are able to

C. are preference to 　　　　　　D. are in accordance with

9. These factors lower overall maintenance requirements by reducing the frequency of adsorbent change out and minimizing <u>downtime</u> – related losses in production.

A. uptime　　B. downturn　　C. downward　　D. shutdown

10. Inexpensive, noncorrosive, nontoxic, chemically inert, high bulk <u>density</u> and no significant volume changes upon adsorption and desorption of water.

A. decomposition 　　　　　　B. suitability

C. concentration 　　　　　　D. dehydration

Ⅲ. Multiple choice：read the four suggested translations and choose the best answer.

1. Water can condense in the pipeline, causing <u>slug flow</u> and possible erosion and corrosion.

A. 段塞流　　B. 漫流　　　C. 游离水　　　D. 结晶体

2. <u>Multistage separators</u> can also be deployed to ensure the reduction of free water that may be present.

A. 多级分离器 　　　　　　B. 往复式压缩机

C. 燃气轮机 　　　　　　　D. 分层离析机

3. The most common of these are <u>liquid desiccant（glycol）dehydration</u>, solid desiccant dehydration, and refrigeration.

A. 化学抑制剂 　　　　　　B. 电解质溶液

C. 天然气的分离器 　　　　D. 液体干燥剂脱水法

4. The first two methods utilize mass transfer of the water molecule into a <u>liquid solvent</u>（glycol solution）or a crystalline structure（dry desiccant）.

A. 溶解液体 　　　　　　　B. 吸收溶液

C. 液体溶剂 　　　　　　　D. 气体干燥

5. TEG is by far the most common liquid desiccant used in natural gas dehydration. It exhibits most of the desirable criteria of <u>commercial suitability</u> listed here.

A. 商业适应 　　　　　　　B. 商业指标

C. 商务活动　　　　　　D. 商务标准

6. TEG is regenerated more easily to a concentration of 98 – 99% in an atmospheric stripper because of its high boiling point and decomposition temperature.

A. 大气清新器　　　　　B. 空气剥离器

C. 常压汽提器　　　　　D. 气压脱模机

7. TEG has an initial theoretical decomposition temperature of 404 ℉, whereas that of diethylene glycol is only 328 ℉.

A. 首个理论分解气温　　B. 初始理论分解温度

C. 初次推理分解温度　　D. 初次假设分解气温

8. The water forms an extremely thin film that is held to the desiccant surface by forces of attraction, but there is no chemical reaction.

A. 极薄的液膜　　　　　B. 特薄的薄膜

C. 特薄的胶片　　　　　D. 很薄的影片

9. Using desiccant dehydrators as alternatives to glycol dehydrators can yield significant economic and environmental benefits, including reduced capital cost, reduced operation and maintenance cost, and minimal VOC and hazardous air pollutants (BTEX).

A. 有毒尾气　　　　　　B. 有毒气体排放

C. 挥发有机化合物　　　D. 有害气体污染

10. The dynamic moisture sorption capacity of a desiccant will depend on a number of factors, such as the relative humidity of the inlet gas, the gas flow rate, the temperature of the adsorption zone, the mesh size of the granule, and the length of service and degree of contamination of the desiccant and not the least on the desiccant itself.

A. 网格大小　　　　　　B. 吸水容量

C. 颗粒大小　　　　　　D. 吸附区大小

IV. **Put the following sentences into Chinese.**

1. Natural, associated, or tail gas usually contains water, in liquid and/or vapor form, at source and/or as a result of sweetening with an aqueous solution.

2. Operating experience and thorough engineering have proved that it is necessary to reduce and control the water content of gas to ensure safe process-

ing and transmission.

3. Using desiccant dehydrators as alternatives to glycol dehydrators can yield significant economic and environmental benefits, including reduced capital cost, reduced operation and maintenance cost, and minimal VOC and hazardous air pollutants (BTEX).

4. Some are good only for dehydrating the gas, whereas others are capable of performing both dehydration and removal of heavy hydrocarbon components.

5. These factors lower overall maintenance requirements by reducing the frequency of adsorbent change out and minimizing downtime – related losses in production.

V. Put the following paragraphs into Chinese.

1. Pipeline drips installed near wellheads and at strategic locations along gathering and trunk lines will eliminate most of the free water lifted from the wells in the gas stream. Multistage separators can also be deployed to ensure the reduction of free water that may be present. However, removal of the water vapor that exists in solution in natural gas requires a more complex treatment. This treatment consists of "dehydrating" the natural gas, which is accomplished by lowering the dew point temperature of the gas at which water vapor will condense from the gas.

2. There are several methods of dehydrating natural gas. The most common of these are liquid desiccant (glycol) dehydration, solid desiccant dehydration, and refrigeration (i. e. , cooling the gas). The first two methods utilize mass transfer of the water molecule into a liquid solvent (glycol solution) or a crystalline structure (dry desiccant). The third method employs cooling to condense the water molecule to the liquid phase with the subsequent injection of inhibitor to prevent hydrate formation. However, the choice of dehydration method is usually between glycol and solid desiccants.

6. 4　Wax Treatment

Guidance to Reading

*Multiphase flow can be severely affected by the deposition of organic solids, usually in the form of **wax crystals**, and their potential to disrupt production due to deposition in the production/transmission systems. The deposits* *cause subsurface and surface equipment **plugging** and **malfunction**, especially when oil mixtures are transported across Arctic regions or through cold oceans. **Wax deposition** leads to more frequent and risky pigging requirements in pipelines. If the wax deposits get too thick, they often reduce the capacity of the pipeline and cause the **pigs** to get stuck. Wax deposition in well tubings and process equipment may lead to more frequent shutdowns and operational problems.*

[中] Text

Wax Deposition

Precipitation of wax from petroleum fluids is considered to be a **thermodynamic molecular saturation phenomenon**. **Paraffin wax** molecules are initially dissolved in a chaotic molecular state in the fluid. At some thermodynamic state the fluid becomes saturated with the wax molecules, which then begin to precipitate. This thermodynamic state is called the onset of wax precipitation or solidification. It is **analogous** to the usual dew point or condensation phenomenon, except that in wax precipitation a solid is precipitating from a liquid, whereas in condensation a liquid is precipitating from a vapor. In wax precipitation, **resin** and **asphaltene micelles** behave like heavy molecules. When their kinetic energy is sufficiently reduced due to cooling, they precipitate out of solution but they are not destroyed. If kinetic energy in the form of heat is supplied to the system, these micelles will **desegregate** and go back into stable suspension and **Brownian motion**.

Wax Deposition Inhibition/Prevention

In oil flow lines, wax deposition occurs by **diffusion** of wax molecules and crystals toward and attachment at the wall. The wax crystals reduce the effective cross – sec tional area of the pipe and increase the pipeline roughness, which results in an increase in pressure drop.

Prevention or inhibition of wax deposition is mainly accomplished by injection of a special class of molecules that interact with paraffin molecules at temperatures above the **cloud point** and influence their crystallization process in a way that diminishes the attraction of the formed crystals toward the wall. The inhibited formed wax crystals are removed from the system by the shear forces caused by the flowing oil.

The **aforementioned** mechanism occurs when the wax crystal forms at the wall. This is the case with most liquid – filled **bare (uninsulated) lines**. The two important requirements are fast heat transfer and fast diffusion of paraffin molecules toward the wall. This is not the case in many situations of flow lines carrying gas/condensates. There is fast cooling but the majority of the gas cools down while in the main flow, thus forming wax crystals with the main flow that would have a tendency to deposit or sediment by gravity in the liquid holdup. Also, the liquid that forms near the **inlet** of the bare line that has not reached its cloud point cools with a velocity that is almost standstill compared to the main flow. Slugging in this case would be beneficial in removing **wax slush** from the line.

Additionally, these flow lines are essentially soaked in methanol to prevent hydrate formation. It just so happens that most (if not all) chemicals with wax crystal modification properties are **incompatible** with **alcohols** and glycols. Hence, the inhibited wax crystals accumulate in the flowline in the liquid hold-up, thus adding to the wax accumulation. This accumulation forms what has been called "wax slush". This wax slush is viscous but it moves and can flow given enough shear force.

It is appropriate at this time to give a brief description of the main two chemical classes used to treat wax deposition. There are three types of wax crystals：

(1) Plate crystals.

(2) Needle crystals.

(3) Mal or amorphous crystals.

Paraffinic oils form plate or needle crystals. **Asphaltenic oils** form primarily mal or **amorphous** crystals. Asphaltenes act as **nucleation sites** for wax crystal growth into mal crystals. Plate crystals look as their name implies, like plates under the microscope. Needle crystals look like needles, and mal crystals are amorphous and generally look like small round spheres. The interaction between crystals and pipe wall increases from mal to needle to plates. Thus, maintaining newly formed wax crystals small and round, i. e. , like mal crystals, is desirable.

The behavior and properties, e. g. , cloud point and **pour point**, of paraffin crystals precipitating from a hydrocarbon can be affected in three ways.

(1)Crystal size modification: modification of the crystal from larger sizes to smaller sizes.

(2)**Nucleation inhibition**: inhibition of the growth rate of the crystal and its ultimate size.

(3)Crystal type or structure modification: modification of the crystal from one type to the other. For instance, modify a crystal from needle to mal type.

A wax crystal modifier works primarily to modify the crystal size. The plate crystals of an *n* – **paraffin** look much smaller under the microscope when precipitating from hydrocarbons inhibited with a wax crystal modifier. Smaller crystals have lower molecular weights and thus higher **solubility** in oil. Furthermore, smaller particles have smaller energy of interaction among themselves and the pipe wall. A crystal modifier interrupts the normal crystal growth of the *n* – paraffins by inserting itself in the crystal, thus stopping its growth.

It is noted that another name for a wax crystal modifier is **pour point depressant** (**PPD**). As the name implies, wax crystal modifiers are very effective at suppres sing the pour points of crude oils because they suppress the wax crystal growth, thus minimizing the strength of their interactions.

A wax **dispersant** may act to inhibit wax nucleation and change the type of wax crystals from plate or needle to mal or amorphous. Inhibited amorphous crystals are smaller and carried much easier by the hydrodynamic drag of the flowing fluid. The presence of asphaltenes and resins facilitates the effect of the dispersant. The dispersant interacts with the asphaltenes and resins and ties them up, thus removing nucleation sites required by wax crystals to grow. A dispersant, when added to the oil at a temperature above the cloud point, has occasionally an additional benefit of suppres sing the cloud point by interacting with and tying up the asphaltenes and resins that tend to come out first. A dispersant tends to usually disperse the wax particles at the water – oil interface.

Both wax crystal modifiers and wax dispersants are useful chemicals that have the ability to diminish wax formation and deposition, although through different mechanisms. Wax dispersants are usually much smaller in molecular weight and size than wax crystal modifiers. Hence, their viscosity and flow properties in general are more favorable in cold applications. Also, some wax dispersants are soluble with alcohols and glycols, thus making them compatible for simultaneous injection. The selection of the wax chemical should be made after careful consideration of the produced hydrocarbon and facilities.

Wax Deposit Remediation

Pigging is an option, but only after very careful consideration of the system's performance to understand the dynamics of the moving pig and continuous removal of the wax cuttings ahead of it. This is a very difficult job. Controlling the bypass of flow around the pig for such a purpose is difficult. Hence, many pigging operations end up in failure with stuck pigs.

The pigging analysis and decision must be left to true experts. Starting a pigging program at the beginning of the life of the system has a better chance of success than at any other time. Even then, excellent monitoring of the system's PI is a must. It is recommended that a short shutdown, chemical soak, and fast start – up be considered first, because it is the safest option.

Words and Expressions

plugging	堵塞
malfunction	故障
pig	清管器
precipitation	析出
thermodynamic	热动力学的
analogous	类似的
resin	树脂
asphaltene	沥青
micelle	微团
desegregate	分散
diffusion	分散
aforementioned	上述的
uninsulated	不保温的
inlet	入口
incompatible	不相溶的
alcohol	酒精
amorphous	无定形的,非晶质,非结晶的
solubility	可溶性
dispersant	分散剂

Phrases and Expressions

wax crystal	蜡状物
wax deposition	结蜡
molecular saturation phenomenon	分子饱和现象
paraffin wax	固体石蜡
Brownian motion	布朗运动
cloud point	浊点
bare line	裸管
wax slush	蜡泥
plate crystal	片状晶体
needle crystal	针状晶体
mal or amorphous crystal	无定形晶体
paraffinic oil	石蜡油
asphaltenic oil	沥青质油
nucleation site	成核位置
nucleation inhibition	成核作用抑制
pour point	倾点
n – paraffin	正烷烃
pour point depressant(PPD)	降凝剂

Language Focus

1. The wax crystals reduce the effective cross – sectional area of the pipe and increase the pipeline roughness, which results in an increase in pressure drop.

(参考译文:结蜡减少了管线的有效输送面积,增大了管壁粗糙度,导致压降增大。)

本句中"and"连接两个并列的动宾短语"reduce the effective cross – sectional area of the pipe"和"increase the pipeline roughness";"which"引导非限制性定语从句"results in an increase in pressure drop",指代前面整句,"results in"是固定搭配,意为"导致,结果是"。

2. It is analogous to the usual dew point or condensation phenomenon, except that in wax precipitation a solid is precipitating from a liquid, whereas in condensation a liquid is precipitating from a vapor.

（参考译文：这种现象和液体凝析现象中的露点是类似的，所不同的是结蜡过程是固体从液体中析出，凝析过程是液体从气体中析出。）

本句中"be analogous to"是固定搭配，意为"与……类似"；"except"作连词，意为"除了……，只……"；"whereas"是连词，意为"然而"，表示对比和转折。

3. There is fast cooling but the majority of the gas cools down while in the main flow, thus forming wax crystals with the main flow that would have a tendency to deposit or sediment by gravity in the liquid holdup.

（参考译文：虽然可以快速冷却，但是大部分气体在主流中流动冷却，因此，和主流液体形成石蜡晶体，由于重力的作用有沉淀和下降的趋势。）

本句中"but"作连词，连接两个并列句，表示转折；"while"在"but"分句中作连词，意为"当……时候"；"thus"作副词，意为"因此"，后面接现在分词短语做伴随状语；"that"引导限制性定语从句，指代"wax crystals with the main flow"。

4. As the name implies, wax crystal modifiers are very effective at suppressing the pour points of crude oils because they suppress the wax crystal growth, thus minimizing the strength of their interactions.

（参考译文：顾名思义，蜡晶改性剂能够抑制蜡晶体的增长，从而有效降低原油的倾点，因此降低它们之间相互作用的强度。）

本句中"as the name implies"意为"顾名思义"，"as"为连词；"because"引导原因状语从句；"thus"后面接现在分词短语，为结果状语。

5. It is recommended that a short shutdown, chemical soak, and fast start – up be considered first, because it is the safest option.

（参考译文：建议首先考虑临时关井、化学剂浸泡，然后快速启动，这是最安全的选择。）

本句中"It is recommended that"意为"有人推荐，建议"，从句用虚拟语气，动词原形前should可以省略，应该是"should be considered"；"because"引导原因状语从句。

🔲 Reinforced Learning

Ⅰ. Answer the following questions for a comprehension of the text.

1. What problems does the deposition of organic solids usually bring?

2. What are the methods of prevention or inhibition of wax deposition?

3. What are the three ways that can affect the behavior and properties of

paraffin crystals precipitating from a hydrocarbon?

    4. What is the primary role of a wax crystal modifier?

    5. What should we consider when pigging is selected?

    II. Multiple choice: choose the correct one from the alternative answers to give the exact meaning of the words.

    1. The <u>deposits</u> also cause subsurface and surface equipment plugging and malfunction, especially when oil mixtures are transported across Arctic regions or through cold oceans.

    A. disposition   B. sediments   C. withdrawal   D. sinkage

    2. It is <u>analogous</u> to the usual dew point or condensation phenomenon, except that in wax precipitation a solid is precipitating from a liquid, whereas in condensation a liquid is precipitating from a vapor.

    A. similar   B. anonymous   C. contrary   D. the same

    3. If kinetic energy in the form of heat is supplied to the system, these micelles will <u>desegregate</u> and go back into stable suspension and Brownian motion.

    A. merge   B. intermingle   C. disperse   D. accelerate

    4. The <u>aforementioned</u> mechanism occurs when the wax crystal forms at the wall.

    A. referred   B. above   C. before   D. below

    5. It just so happens that most (if not all) chemicals with wax crystal modification properties are <u>incompatible</u> with alcohols and glycols.

    A. unsuitable   B. incomparable

    C. impenetrable   D. dissolved

    6. Thus, maintaining newly formed wax crystals small and round, i. e. , like mal crystals, is <u>desirable</u>.

    A. agreeable   B. amiable   C. suitable   D. available

    7. <u>Furthermore</u>, smaller particles have smaller energy of interaction among themselves and the pipe wall.

    A. Moreover   B. Thus   C. Therefore   D. However

    8. Hence, their vis cos ity and flow properties <u>in general</u> are more favorable in cold applications.

    A. as a result   B. in a way   C. as a rule   D. on occasions

    9. Pigging is an <u>option</u>, but only after very careful consideration of the

system's performance to understand the dynamics of the moving pig and continuous removal of the wax cuttings ahead of it.

A. technique   B. choice   C. process   D. step

10. Hence, many pigging operations end up in failure with stuck pigs.

A. result from   B. lead to   C. give rise to   D. end up with

III. Multiple choice: read the four suggested translations and choose the best answer.

1. Precipitation of wax from petroleum fluids is considered to be a thermodynamic molecular saturation phenomenon.

A. 降蜡   B. 结蜡   C. 沉蜡   D. 析蜡

2. This thermodynamic state is called the onset of wax precipitation or solidification.

A. 蜡沉降或蜡凝固   B. 结蜡或固蜡

C. 蜡沉降或蜡浓缩   D. 析蜡点或凝固点

3. In wax precipitation, resin and asphaltene micelles behave like heavy molecules.

A. 重微粒   B. 重摩尔   C. 大微粒   D. 大分子

4. The behavior and properties, e. g. , cloud point and pour point, of paraffin crystals precipitating from a hydrocarbon can be affected in three ways.

A. 行为与性能   B. 特性与属性

C. 特点与内容   D. 行为与特性

5. The plate crystals of an *n* − paraffin look much smaller under the microscope when precipitating from hydrocarbons inhibited with a wax crystal modifier.

A. 蜡晶体修改剂   B. 蜡晶体修饰剂

C. 蜡晶体改良剂   D. 蜡晶体调节器

6. Smaller crystals have lower molecular weights and thus higher solubility in oil.

A. 解决度   B. 溶解量   C. 可溶性   D. 分解度

7. A wax dispersant may act to inhibit wax nucleation and change the type of wax crystals from plate or needle to mal or amorphous.

A. 分散剂   B. 降凝剂   C. 改良剂   D. 改性剂

8. Both wax crystal modifiers and wax dispersants are useful chemicals that have the ability to diminish wax formation and deposition, although through different mechanisms.

A. 机制　　　B. 机理　　　C. 机构　　　D. 机械

9. The pigging analysis and decision must be left to true experts.

A. 教授　　　B. 人才　　　C. 工程师　　　D. 专家

10. It is recommended that a short shutdown, chemical soak, and fast start — up be considered first, because it is the safest option.

A. 化学学习　　　　　　　B. 化学渗透
C. 化学剂浸泡　　　　　　D. 吸入化学剂

Ⅳ. Put the following sentences into Chinese.

1. Multiphase flow can be severely affected by the deposition of organic solids, usually in the form of wax crystals, and their potential to disrupt production due to deposition in the production/transmission systems.

2. Prevention or inhibition of wax deposition is mainly accomplished by injection of a special class of molecules that interact with paraffin molecules at temperatures above the cloud point and influence their crystallization process in a way that diminishes the attraction of the formed crystals toward the wall.

3. It is appropriate at this time to give a brief description of the main two chemical classes used to treat wax deposition.

4. The behavior and properties, e. g. , cloud point and pour point, of paraffin crystals precipitating from a hydrocarbon can be affected in three ways.

5. Pigging is an option, but only after very careful consideration of the system's performance to understand the dynamics of the moving pig and continuous removal of the wax cuttings ahead of it.

Ⅴ. Put the following paragraphs into Chinese.

1. Precipitation of wax from petroleum fluids is considered to be a thermodynamic molecular saturation phenomenon. Paraffin wax molecules are initially dissolved in a chaotic molecular state in the fluid. At some thermodynamic state the fluid becomes saturated with the wax molecules, which then begin to precipitate. This thermodynamic state is called the onset of wax precipitation or solidification. It is analogous to the usual dew point or condensation phenomenon, except that in wax precipitation a solid is precipitating from a liquid, whereas in condensation a liquid is precipitating from a vapor.

2. Both wax crystal modifiers and wax dispersants are useful chemicals that have the ability to diminish wax formation and deposition, although

through different mechanisms. Wax dispersants are usually much smaller in molecular weight and size than wax crystal modifiers. Hence, their vis cos ity and flow properties in general are more favorable in cold applications. Also, some wax dispersants are soluble with alcohols and glycols, thus making them compatible for simultaneous injection.

6.5　Natural Gas Compression

🔲 Guidance to Reading

*Natural gas compression is used to increase pipeline – operating pressures, which can bring about the benefits of the ability to transmit larger volumes of gas through a given size of pipeline, lower **transmission** losses due to friction, and the capability to transmit gas over long distances without additional boosting stations. Two basic types of compressors are usually used: reciprocating and **centrifugal compressors**. The former are usually driven by either electric motors or gas engines, whereas the latter use gas turbines or electric motors as drivers. Both have their own advantages and the selection can be made based on some principles mentioned in this paper.*

🔲 Text

"Compression" is used in all aspects of the natural gas industry, including gas lift, **reinjection** of gas for pressure **maintenance**, gas gathering, gas proces sing operations (circulation of gas through the process or system), transmission and **distribution** systems, and reducing the gas volume for shipment by tankers or for storage. In recent years, there has been a trend toward increa sing pipe-line – operating pressures. The benefits of operating at higher pressures include the ability to transmit larger volumes of gas through a given size of pipeline, lower transmission losses due to friction, and the capability to transmit gas over long distances without additional boosting stations. In gas transmission, two basic types of compressors are used: reciprocating and centrifugal compressors. **Reciprocating compressors** are usually driven by either electric motors or gas engines, whereas centrifugal compressors use gas turbines or electric motors as drivers.

The key variables for equipment selections are **life cycle** cost, **capital** cost, maintenance costs, including overhaul and spare parts, fuel, or energy costs. The units level of utilization, as well as demand fluctuations, plays an important role. While both gas engines and gas turbines can use pipeline gas as a fuel, an electric motor has to rely on the availability of electric power. Due to the number of variables involved, the task of choo sing the optimum driver can be quite involved, and a comparison between the different types of drivers should be done before a final selection is made. An economic feasibility study is of fundamental importance to determine the best selection for the economic life of a project. Furthermore, it must be decided whether the compression task should be divided into multiple compressor trains, operating **in series** or **in parallel**.

Reciprocating Compressors

A reciprocating compressor is a positive displacement machine in which the compres sing and displacing element is a **piston** moving linearly within a **cylinder**. The reciprocating compressor uses automatic spring – loaded valves that open when the proper differential pressure exists across the valve. When the suction valve is open, gas will flow into the cylinder and compressed. The flow to and from reciprocating compressors is subject to significant pressure fluctuations due to the reciprocating compression process. Therefore, **pulsation dampeners** have to be installed upstream and downstream of the compressor to avoid damages to other equipment. The pressure losses (several percent of the static flow pressure) of these dampeners have to be accounted for in the station design.

Reciprocating compressors are widely utilized in the gas proces sing industries because they are flexible in throughput and discharge pressure range. Reciprocating compressors are classified as either "high speed" or "slow speed". Typically, high – speed compressors operate at speeds of 900 to 1200 rpm and slow – speed units at speeds of 200 to 600 rpm. High – speed units are normally "separable", i. e. , the compressor frame and driver are separated by a **coupling** or **gearbox**. For an "integral" unit, power cylinders are mounted on the same frame as the compressor cylinders, and power pistons are attached to the same drive shaft as the compressor cylinders. Low – speed units are typically integral in design.

Centrifugal Compressors

A centrifugal compressor stage is defined as one **impeller**, with the subsequent **diffuser** and (if applicable) return channel. A compressor body may hold one or several (up to 8 or 10) stages. A compressor train may consist of one or multiple compressor bodies. It sometimes also includes a gearbox. Pipeline compressors are typically single body trains, with one or two stages.

The different working principles cause differences in the operating characteristics of the centrifugal compressors compared to those of the reciprocating unit. Centrifugal compressors are used in a wide variety of applications in chemical plants, refineries, onshore and offshore gas lift and gas injection applications, gas gathering, and in the transmission of natural gas. Centrifugal compressors can be used for outlet pressures as high as 10,000 psia, thus overlapping with reciprocating compressors over a portion of the flow rate/pressure domain. Centrifugal compressors are usually either turbine or electric motor driven. Typical operating speeds for centrifugal compressors in gas transmission applications are about 14000 rpm for 5000 – hp units and 8000 rpm for 20000 – hp units.

Comparison Between Compressors

Differences between reciprocating and centrifugal compressors are summarized as follow.

Advantages of a reciprocating compressor over a centrifugal machine include

(1) Ideal for low volume flow and high – pressure ratios.

(2) High efficiency at high – pressure ratios.

(3) Relatively low capital cost in small units (less than 3000 hp).

(4) Less sensitive to changes in composition and density.

Advantages of a centrifugal compressor over a reciprocating machine include

(1) Ideal for high volume flow and low head.

(2) Simple construction with only one moving part.

(3) High efficiency over normal operating range.

(4) Low maintenance cost and high availability.

(5) Greater volume capacity per unit of plot area.

(6) No **vibrations** and **pulsations** generated.

Compressor Selection

The design philosophy for choo sing a compressor should include the following considerations.

(1) Good efficiency over a wide range of operating conditions.

(2) Maximum flexibility of **configuration**.

(3) Low maintenance cost.

(4) Low life cycle cost.

(5) Acceptable capital cost.

(6) High availability.

However, additional requirements and features will depend on each project and on specific experiences of the pipeline operator. In fact, compressor selection consists of the purchaser defining the operating parameters for which the machine will be designed. The "process design parameters" that specify a selection are as follows.

(1) Flow rate.

(2) Gas composition.

(3) Inlet pressure and temperature.

(4) Outlet pressure.

(5) Train arrangement:

① For centrifugal compressors: series, parallel, multiple bodies, multiple sec tions, intercooling, etc.

② For reciprocating compressors: number of cylinders, cooling, and flow control strategy.

(6) Number of units.

In many cases, the decision whether to use a reciprocating compressor or a centrifugal compressor, as well as the type of driver, will already have been made based on operator strategy, emissions requirements, general life cycle cost assumptions, and so on. However, a **hydraulic analysis** should be made for each compressor selection to ensure the best choice. In fact, compressor selection can be made for an operating point that will be the most likely or most frequent operating point of the machine. Selections based on a single operating point have to be evaluated carefully to provide sufficient speed margin (typically 3 to 10%) and surge margin to cover other, potentially important situations. A compressor performance map can be generated based on the selection and is used to evaluate the compressor for other operating conditions by determining the head and flow required for these other operating conditions. In many applications, multiple operating points are available, e. g. , based on hydraulic pipeline studies or reservoir studies. Some of these points may be frequent operating points, while some may just occur during upset conditions. With this knowledge, the selection can be optimized for a desired target, such as lowest fuel consumption.

Words and Expressions

transmission	输送
reinjection	回注
maintenance	保持
distribution	分配
piston	活塞
cylinder	气缸
coupling	联轴器
gearbox	变速箱
impeller	推进器
diffuser	扩散器
vibration	震动
pulsation	脉动
configuration	配置

Phrases and Expressions

centrifugal compressor	离心式压缩机
reciprocating compressor	往复式压缩机
life cycle	寿命周期
capital cost	资本成本
in series	串联
in parallel	并联
pulsation dampener	压力缓冲器
hydraulic analysis	水力分析

Language Focus

1. Reciprocating compressors are usually driven by either electric motors or gas engines, whereas centrifugal compressors use gas turbines or electric mo-

tors as drivers.

（参考译文：往复式压缩机通常是用电动机或燃气发动机作动力，而离心式压缩机则是用燃气轮机或电动机来驱动。）

本句中"whereas"作连词，意为"而"，表示对比。

2. In fact, compressor selection consists of the purchaser defining the operating parameters for which the machine will be designed.

（参考译文：实际上，压缩机选择主要是根据购买者提出的操作参数对机器进行设计。）

本句中"in fact"是副词短语，意为"事实上"；谓语"consists of"后面跟了一个独立主格结构作宾语，独立主格中逻辑主语是"the purchaser"，现在分词短语"defining the operating parameters. . . "作逻辑谓语；"which"引导限制性定语从句，指代"the operating parameters"。

🔁 Reinforced Learning

Ⅰ. Answer the following questions for a comprehension of the text.

1. What are the two basic types of compressors in gas transmission?

2. What are the key variables for equipment selections?

3. What are the differences between reciprocating compressors and centrifugal compressors?

4. What are the basic principles for choo sing a compressor?

5. What other factors should be considered for compressor selection?

Ⅱ. Multiple choice: choose the correct one from the alternative answers to give the exact meaning of the words.

1. The units level of utilization, as well as demand fluctuations, plays an important role.

    A. variation   B. maintenance   C. compression   D. transmission

2. While both gas engines and gas turbines can use pipeline gas as a fuel, an electric motor has to rely on the availability of electric power.

    A. accessibility   B. capability

    C. circulation   D. selection

3. An economic feasibility study is of fundamental importance to determine the best selection for the economic life of a project.

A. finance   B. presentation   C. utilization   D. feasibleness

4. The pressure losses (several percent of the static flow pressure) of these dampeners have to be <u>accounted</u> for in the station design.

A. displaced   B. considered   C. loaded   D. distributed

5. For an "integral" unit, power cylinders are mounted on the same frame as the compressor cylinders, and power pistons are <u>attached</u> to the same drive shaft as the compressor cylinders.

A. framed   B. mounted   C. affiliated   D. decorated

6. A centrifugal compressor stage is defined as one <u>impeller</u>, with the subsequent diffuser and (if applicable) return channel.

A. gearbox   B. rotor   C. channel   D. compressor

7. The different working principles cause differences in the operating characteristics of the centrifugal compressors compared to those of the reciprocating <u>unit</u>.

A. diffuser   B. cylinder   C. dampeners   D. compressors

8. Centrifugal compressors are used <u>in a wide variety of</u> applications in chemical plants, refineries, onshore and offshore gas lift and gas injection applications, gas gathering, and in the transmission of natural gas.

A. some   B. many   C. various   D. a few

9. Centrifugal compressors can be used for outlet pressures as high as 10000 psia, thus overlapping with reciprocating compressors over a portion of the flow rate/pressure <u>domain</u>.

A. design   B. channel   C. range   D. equipment

10. The design <u>philosophy</u> for choosing a compressor should include the following considerations.

A. concept   B. scheme   C. plan   D. strategy

III. Multiple choice:read the four suggested translations and choose the best answer.

1. "Compression" is used in all aspects of the natural gas industry, including gas lift, <u>reinjection of gas</u> for pressure maintenance, gas gathering, gas proces sing operations (circulation of gas through the process or system),

transmission and distribution systems, and reducing the gas volume for shipment by tankers or for storage.

    A. 储存气体   B. 天然气输送

    C. 气体回注   D. 天然气处理

    2. The benefits of operating at higher pressures include the ability to transmit larger volumes of gas through a given size of pipeline, lower transmission losses due to friction, and the capability to transmit gas over long distances without additional boosting stations.

    A. 增压站   B. 扩散器   C. 压缩机   D. 弹簧阀门

    3. In gas transmission, two basic types of compressors are used: reciprocating and centrifugal compressors.

    A. 天然气储集和处理生产厂   B. 往复式压缩机和离心式压缩机

    C. 往复式压缩机和分离压缩机   D. 低速压缩机和输送管线

    4. The key variables for equipment selections are life cycle cost, capital cost, maintenance costs, including overhaul and spare parts, fuel, or energy costs.

    A. 生活成本   B. 生命循环   C. 寿命周期   D. 人寿保险

    5. Furthermore, it must be decided whether the compression task should be divided into multiple compressor trains, operating in series or in parallel.

    A. 连接或平行   B. 系列或集合

    C. 顺联或并行   D. 串联或并联

    6. Therefore, pulsation dampeners have to be installed upstream and downstream of the compressor to avoid damages to other equipment.

    A. 压力缓冲器   B. 高速压缩机

    C. 低速压缩机   D. 压缩机气缸

    7. Reciprocating compressors are widely utilized in the gas processing industries because they are flexible in throughput and discharge pressure range.

    A. 静态流压部分   B. 压力损失

    C. 释放压力排序   D. 排出压力范围

    8. High-speed units are normally "separable", i.e., the compressor frame and driver are separated by a coupling or gearbox.

    A. 扩散器和回流腔   B. 联轴器或变速箱

    C. 压缩机身和驱动装置   D. 动力气缸和压缩机气缸

9. However, additional requirements and features will depend on each project and on specific experiences of the pipeline operator.

A. 管线接线员   B. 油管经营者

C. 管线铺设者   D. 管道操作者

10. In fact, compressor selection consists of the purchaser defining the operating parameters for which the machine will be designed.

A. 额外要求   B. 经营业绩   C. 操作参数   D. 营业参量

Ⅳ. Put the following sentences into Chinese.

1. The key variables for equipment selections are life cycle cost, capital cost, maintenance costs, including overhaul and spare parts, fuel, or energy costs.

2. An economic feasibility study is of fundamental importance to determine the best selection for the economic life of a project.

3. For an "integral" unit, power cylinders are mounted on the same frame as the compressor cylinders, and power pistons are attached to the same drive shaft as the compressor cylinders.

4. The different working principles cause differences in the operating characteristics of the centrifugal compressors compared to those of the reciprocating unit.

5. However, additional requirements and features will depend on each project and on specific experiences of the pipeline operator.

Ⅴ. Put the following paragraphs into Chinese.

1. "Compression" is used in all aspects of the natural gas industry, including gas lift, reinjection of gas, gas gathering, gas proces sing operations, transmission and distribution systems, and reducing the gas volume for shipment by tankers or for storage. In recent years, there has been a trend toward increasing pipeline – operating pressures. The benefits of operating at higher pressures include the ability to transmit larger volumes of gas through a given size of pipeline, lower transmission losses due to friction, and the capability to transmit gas over long distances without additional boosting stations.

2. In many cases, the decision whether to use a reciprocating compressor or a centrifugal compressor, as well as the type of driver, will already have been made based on operator strategy, emissions requirements, general life cy-

cle cost assumptions, and so on. However, a hydraulic analysis should be made for each compressor selection to ensure the best choice. In fact, compressor selection can be made for an operating point that will be the most likely or most frequent operating point of the machine. Selections based on a single operating point have to be evaluated carefully to provide sufficient speed margin and surge margin to cover other, potentially important situations.

Chinese Translation and Key to Exercises

第1章　油气开采与分离

1.1　油气地面处理简介

🔲 导语

通常情况下,油井采出的是油、气、水及各种杂质的混合流体,必须分离和处理。采油设施可以加工并处理这些流体,把主要成分分离出来投放市场,将废弃物妥善处理以保护环境。分离器设计形式多样。此外,油井必须备有测试及计量设备,以便对每一油井天然气、原油及水的产量正确进行配产。天然气销售合同要求天然气脱出大部分水蒸气,防止水合物形成,增加管道运输能力。

🔲 课文

油井采出的流体一般是油、气和水混合在一起。采油设施将从井下采出的液体分离成三种成分(即通常所说的原油、天然气和水三相),并将该三相流体分别加工成为市场销售产品,或将其处理成为环境保护条例所允许的排放物。在称为"分离器"的机械设备中,气体会从液体中闪蒸出来,"游离水"会从原油中分离出来。经过上述处理工序后,原油会损失其中不少轻烃,而成为一种挥发度(蒸气压)符合市场销售条件且性能稳定的原油。分离器按其形状可分为立式及卧式两种。

分离出来的天然气应经压缩及其他处理后再销售。通常是采用引擎驱动的往复式压缩机进行压缩。在大型设施上或在增压作业时,则采用燃气轮机驱动的离心式压缩机,也可成组地采用大型往复式压缩机。

通常,分离出来的天然气含有大量的水蒸气,故应脱水,使其含水量在允许范围内(一般是低于 $7lb/10^6ft^3$)。通常是采用乙二醇脱水。干乙二醇用泵送至大型立式接触塔内,在该塔内将天然气中的水蒸气吸收。湿乙二醇从分离器中流出至大型卧式重沸器内,在重沸器中加热,并蒸发掉水分。

在某些油气产区,为了降低烃的露点,可能需要去掉些重烃成分。天然

气中可能含有诸如 H_2S 及 CO_2 等污染物,其含量可能大于天然气销售规定允许范围。若遇到这种情况,就需要附加处理设备净化天然气。

从分离器中出来的原油及乳化物必须脱水处理。多数石油公司对原油中可能残留的底部沉淀物及水分(BS+W)的最大含量均有规定。不同油气产区对该含量规定不同,通常在 0.5% ~3% 之间。某些炼油厂对原油中的含盐量作出限定。这样,就需要用淡水对原油进行数级脱盐处理,而且,随后还需要对原油作脱水处理。通常,每千桶原油中的含盐量限定在10~25lb范围内。

采油过程中采出的水在大多数海洋采油平台上排入海中,在陆地油矿上则排入池中蒸发掉。通常,可将采出水排放至处理矿产水的污水井中,或作为回注水使用。从分离器中出来的水必须予以处理,以去掉其中所含的少量油。拟排放至污水井的水必须经过过滤设备,将水中的固相颗粒滤掉。

随同井内流体而带出的固相物也必须予以分离、净化,处理后的排放物应符合环保条例的规定。处理装置可为沉降池或沉降罐、旋流分离器、过滤器等。

油井必须备有测试及计量设备,以便对每一油井天然气、原油及水的产量正确进行配产。这样做不仅是为了计量,也是为了在油田行将枯竭之时对油层进行评价研究。

矿场自动接收、取样、计量、传输系统(LACT)

在大型采油厂,原油经 LACT 输出后卖给用户。该系统的设计符合 API 标准,以及根据买方要求而增加的计量及取样标准。接收到的油量取决于它的相对密度、BS+W 值和体积。因此,LACT 系统不仅应准确计量原油量,而且应连续检测 BS+W 的含量,并能从该系统中取出有代表性的原油油样,以便测出原油的相对密度和 BS+W 值。

图 1.1.1 表示常规的 LACT 系统组件及流程。原油首先要流经过滤器/天然气逸出器以保护流量计不至损坏,并保证油流中不含天然气。垂向安装针对底部沉淀物及水含量的自动探测器。当 BS+W 含量超出销售合同规定的标准时,该探测器自动启动而关闭分流阀,使油流不再继续进入LACT 系统,并且将油流反输回来,再次处理。某些销售合同要求 BS+W 超标时,探测器发出警报即可,以便使操作者进行正确的人工处理。如果想获取油流平均质量的真实读数,则 BS+W 探测器应安装在垂向流动处。

取样器从流量计接收信号,以确保油流变化时所取油样与总油流成比例,具有代表性。油样筒备有混合泵,油流在筒内搅拌,以便在取油样之前使该油液搅拌均匀,这就是通过少量油样可计量出 BS+W 值及原油相对密度的原因。

该油流然后流经正排流量计进行计量。在大型采油装置上,如图 1.1.1 所示,流量计的标定仪是作为永久设备装在 LACT 橇装座上的,或者哪里需作标定时就将流量计标定仪带至现场使用。

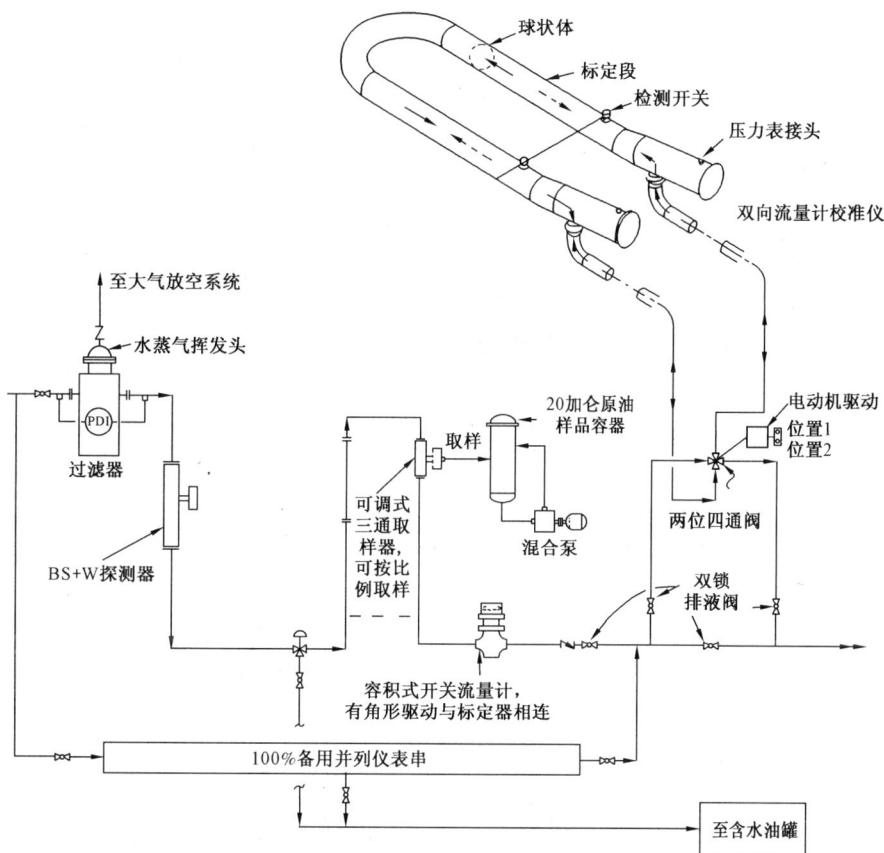

图 1.1.1 常规矿场自动接受、取样、计量、转输系统流程

天然气脱水器

由于天然气在输气及配气过程中会冷凝,要防止其中的水蒸气形成水合物及其引起的腐蚀后患,大多数天然气销售合同都规定,要从天然气中脱出水蒸气。此外,天然气脱水后还可增加管道的输气量。

美国南部的天然气销售合同大都规定:天然气中的含水量要减至 $7lb/10^6ft^3$。在寒冷地区,销售合同通常规定:含水量要低于 $3 \sim 5lb/10^6ft^3$。通常多采用下列方法来使天然气脱水:

(1)将天然气冷却到能生成水合物的温度,并将所生成的水予以分离。允许高含水($\pm30lb/10^6ft^3$)的天然气才能采取此法。

(2)当水合物生成时,可采用低温换热装置来融化水合物。为能有效运行,要求低温换热装置入口压力大于 2500psi。虽然该装置在过去通用,但由于当油井自喷油压下降时有可能冻结,并且在入口压力较低时无法运行,故未得到推广。

(3)使天然气与固态 $CaCl_2$ 层接触。$CaCl_2$ 会吸收天然气中的水分,但 $CaCl_2$ 吸潮后不能再生,且腐蚀性很强。

(4)使用活性矾土、硅胶或分子筛等能再生的固态干燥剂。当然,这些都是相当昂贵的,不过能将天然气中的含水量降到很低。因此,这些干燥剂往往在天然气的低温处理场的入口管上使用,但在采油装置上尚未推广使用。

(5)使用诸如甲醇或乙二醇等不能再生的液态干燥剂。它们费用相当低廉。采用甲醇可降低气井管线中生成水合物的温度,并防止生成水合物阻塞管线。

(6)采用能再生的醇类液体作为干燥剂。这是天然气脱水系统中常用的一种干燥剂。

图 1.1.2 表示常规醇类泡罩接触塔的工作流程。含水天然气由塔的下部进入,并通过层层泡罩由下而上流动。无水醇类进入塔顶,并由每层塔盘的溢流堰溢出,流入下一塔盘。通常大多使用6~8层塔盘。泡罩的作用是确保由下而上的天然气流分散窜至每个小泡罩内,使其最大限度地增大与醇类的接触面。

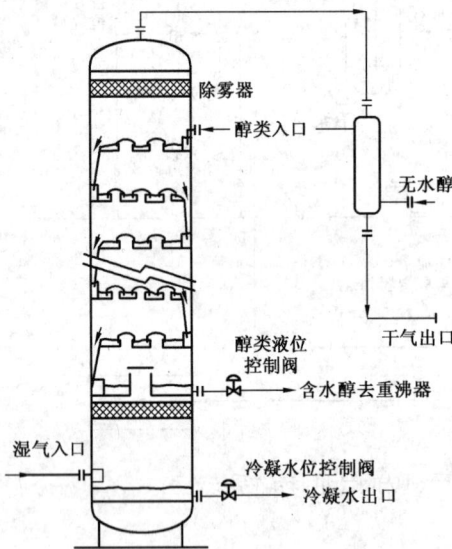

图 1.1.2 常用醇类接触塔

无水醇类进入接触塔之前,要经流出的天然气冷却,以减少进塔时的蒸发损耗。含水醇类从塔的下部流出后,通过换热器,经过天然气的分离器、过滤器进入重沸器。醇类在重沸器内加热至足够高的温度,使水分变成水蒸气逸出。脱水后的醇类再泵入接触塔。

Key to Exercises

Ⅰ.1. The job of a production facility is to separate the well stream into three components, typically called "phases" (oil, gas, and water), and process these phases into some marketable product(s) or dispose of them in an environmentally acceptable manner.

2. If contaminants such as H_2S and CO_2 are at levels higher than those acceptable to the gas purchaser, then additional equipment will be necessary to sweeten the gas.

3. Because gas, oil, and water production can be properly allocated to each well. This is necessary not only for accounting purposes but also to perform reservoir studies as the field is depleted.

4. An automatic BS + W probe is mounted in a vertical run. When BS + W exceeds the sales contract quality this probe automatically actuates the diverter valve, which blocks the liquid from going further in the LACT unit and sends it back to the process for further treating. Some sales contracts allow for the BS + W probe to merely sound a warning so that the operators can manually take corrective action.

5. Most sales contracts in the southern United States call for reducing the water content in the gas to less than 7lb/MMscf.

Ⅱ. 1. D 2. A 3. C 4. A 5. B 6. A 7. C 8. D 9. B 10. C

Ⅲ. 1. B 2. C 3. B 4. D 5. A 6. D 7. A 8. C 9. A 10. C

Ⅳ. 1. 采油设施将从井下采出的液体分离成三种成分(即通常所说的原油、天然气和水三相),并将该三相流体分别加工成为市场销售产品,或将其处理成为环境保护条例所允许的排放物。

2. 分离器按其形状可分为立式及卧式两种。

3. 在大型采油厂,原油经 LACT 输出后卖给用户。该系统的设计符合 API 标准,以及根据买方要求而增加的计量及取样标准。

4. 取样器从流量计接收信号,以确保油流变化时所取油样与总油流成比例,具有代表性。

5. 虽然该装置在过去通用,但由于当油井自喷油压下降时有可能冻结,并且在入口压力较低时无法运行,故未得到推广。

Ⅴ. 1. 采油过程中采出的水在大多数海洋采油平台上排入海中,在陆地油矿上则排入池中蒸发掉。通常,可将采出水排放至处理矿产水的污水井中,或作为回注水使用。从分离器中出来的水必须予以处理,以去掉其中所含的少量油。拟排放至污水井的水必须经过过滤设备,将水中的固相颗粒滤掉。

2. 由于天然气在输气及配气过程中会冷凝,要防止其中的水蒸气形成水合物及其引起的腐蚀后患,大多数天然气销售合同都规定,要从天然气中脱出水蒸气。此外,天然气脱水后还可增加管道的输气量。

1.2 地面生产系统的主要设备

导语

在油气生产中,通常油流是从两口以上的油井混合流入一个处理中心的,因此有必要安装管汇,以便使任一油井所采油流能进入任一容器或其他采油测试系统中去。多级分离可以控制挥发性成分的分离,目的是把水分离出去,获得最多液体回收量和稳定的油气。每一级之间都要有最低压力差,以使压力水平控制良好。

课文

井口装置及管汇

地面油气生产系统的流程是从油井井口开始的。井口至少装有一个油嘴;否则,该油井就不是自喷井,而是抽油井。油井自喷油压与第一个分离器工作压力之间的大部分压降是在油流通过油嘴时产生的。油嘴的开孔直径决定了油流流量,因为油流的上游压力主要取决于油井的自喷油压,而下游压力主要取决于采油过程中第一个分离器上的压力控制阀的压力大小。高压井口最好安装一个固定油嘴,并串联一个可调油嘴。若可调油嘴失控,则固定油嘴可取而代之,并将采出的油流量控制在规定范围内。

在海上采油设备中及其他高风险情况下,采油井口上必须装有紧急自动切断阀(按美国联邦政府法律规定必须如此)。在所有井口设备上,必须安装截断阀,目的是当出油管线较长时能对油嘴进行检修。

只要油流是从两口以上的油井混合流入一个处理中心的,都有必要安装管汇,以便使任一油井所采油流能进入任一容器或其他采油测试系统中去。

初级分离器的压力

由于采出的流体具有多组分特性,因此,初次分离过程中油流压力越高,则分离器中所能获得的液流就越多。该液流中含有一些轻质组分,轻质组分流入分离器下游的存液罐内并逐渐蒸发。

若初次分离过程中的压力太高,则分离器中呈液相的轻质组分也较多,同时在存液罐内该液相会变为气相而蒸发损失掉。若分离器中的压力太低,轻质组分不能稳定于液态中,会变为气相并蒸发而损失掉。

多级分离

在简单的单级分离工艺中,初级分离器中的液体被闪蒸,并且从分离器中出来的液体再次在存液罐中闪蒸,按照惯例,一般不认为存液罐是分离过程中的一个分离级次,虽然它的确有这种作用。图1.2.1表示的是三级分离工艺。液体首先在原有压力下闪蒸,接着在连续低压条件下经过二次闪蒸,最后再进入存液罐。

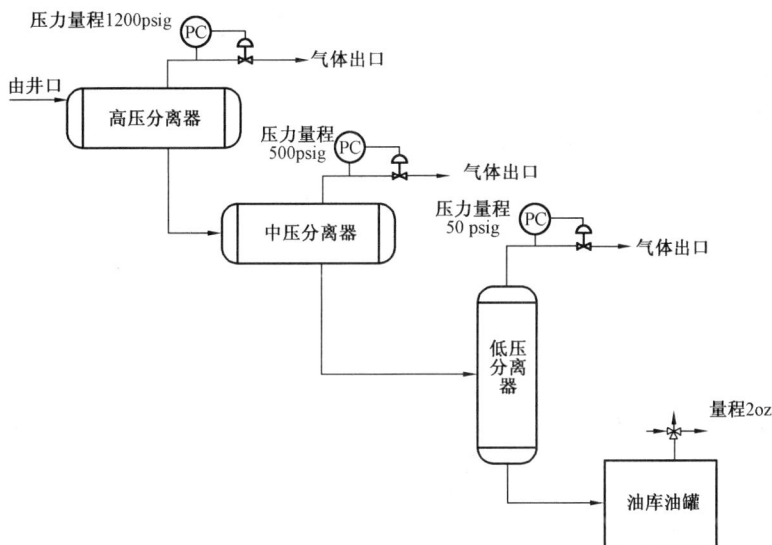

图1.2.1 多级分离

由于采出的流体具有多组分特性,因此由闪蒸计算过程得知:初级分离后,分离次数越多,越多的轻组分就会稳定在液相中。也就是说,在多级分离的工艺条件下,能闪蒸的轻烃分子在相当高压条件下会分离出来,使每级分离过程中的中间烃类分压处于较低状态。若分离级数无限增加,只要轻烃分子一形成,则轻烃分子就会分离出来,并且在每级分离的过程中,中间组分的分压增至最大。当某些气体在高压条件下被捕集时,则通过多级分离所需压缩机的功率也可减小。

分离级数的选择

原油处理过程中分离级数增加得越多,所能回收的液体量就越少。因分离级数增加而相应减少的回收量应能抵得上安装分离器和管路的成本、占用空地的成本及压缩机组的成本。很明显,每一设备均有最佳分离级数。大多数情况下,由于油井和油井的情况不同,而且同一口油井的油流压力也会随时间的推移而下降,因此,要确定最佳分离级数是很困难的。

自喷油压各异的油田

讨论此问题时应注意,在同一油田上的所有油井是在自喷油压大致相同的条件下产油,并且是为了增大产液量及减少压缩机的功率而采用多级分离装置。一般情况下,由于自喷油压不同,故采用多级分离装置。自喷油压的大小可能因油井所在的油层条件不同而各异,或者即使在同一油层但产水量不同。通过配置管汇及初级分离器的不同工作压力,不仅有利于高压井液的多级分离,也会保存油层能量。高压井可将油气输至用户而无需加压;而低压自喷井可将油气输至需要少量加压的系统中。

两相分离器与三相分离器的对比

在我们所举实例中,高压及中等压力等级的分离器均是两相分离器,而低压分离器则是三相分离器。这就是所谓的"游离水脱除器",其设计可使游离水与原油及其乳化物分离,同时可使天然气从液体中分离出来。选用两相分离器还是三相分离器,这要根据油井的流动特性而定。若高压井内出现大量水,而高压分离器具有三相分离功能,则此时可减小其他分离器的尺寸。

原油处理

海上采油装置中的原油处理设备大多做成立式或卧式的,图1.2.2是原油处理设备图。该设备中有一气体覆盖层,其目的是保证原油处理设备中总是有足够的压力,从而使脱出的水能流入污水处理设备中去。

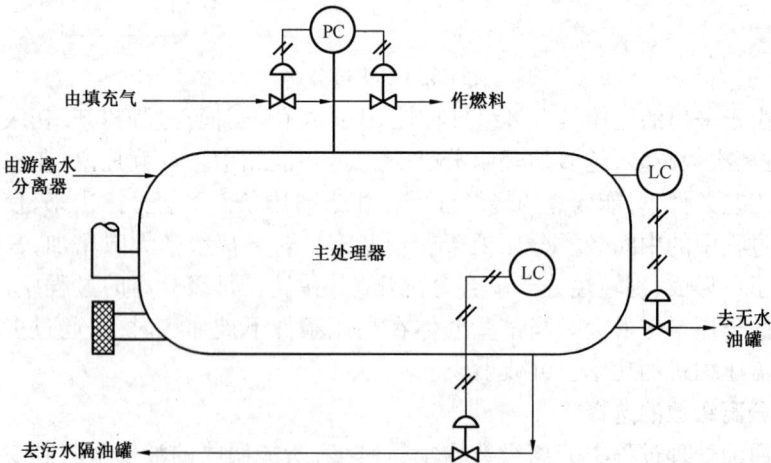

图1.2.2 主处理器

陆地采油设备中,原油是在大型"沉降罐"中进行处理的,如图1.2.3中所示。为了能保证系统内的正压并脱去氧气,该系统中所有罐内必须装有带阻火器的压力/真空阀及气体覆盖层。这样做有助于防腐,消除事故隐患并减少轻烃挥发。图 1.2.4 表示了常规压力/真空阀。罐内压力可将阀内的加重盘顶起,适当泄放罐内的天然气。若由于罐内某一层气罩失效,不能维持稍高正压而出现真空,则外部更高的压力会顶起另一圆盘,使空气进入。虽然我们希望排除空气,但宁可使少许可控量的空气进入罐,也不愿罐体因负压而被压扁。

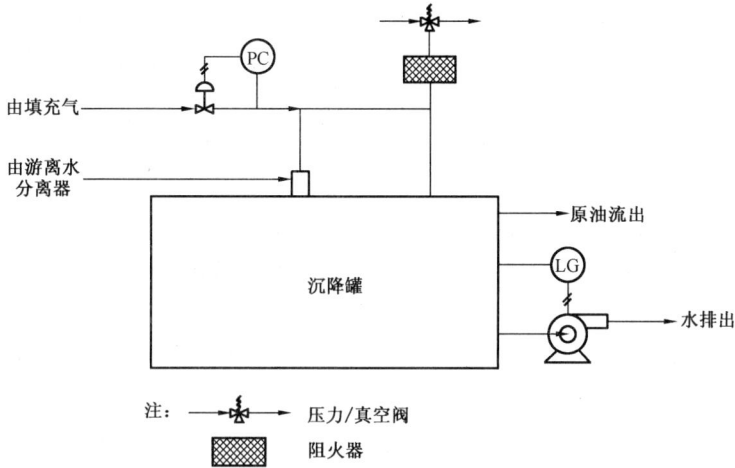

注: 压力/真空阀

阻火器

图 1.2.3　沉降罐

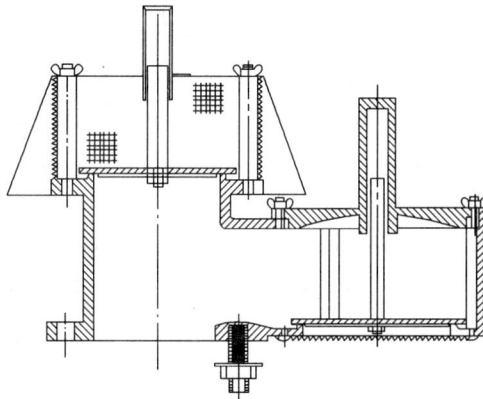

图 1.2.4　常用压力/真空阀

油水分离器(沉降罐)表面的原油可以撇去,而水在含水区,由油水界面控制器及排水阀控制。必须指出,由于原油流出口对液体体积有所限定,故油水分离器不能当作缓冲罐使用。

油流经过处理罐或油水分离器之后进入缓冲罐,再由缓冲罐流入油轮或油罐车,或泵入管道。

⌂ Key to Exercises

Ⅰ. 1. It has at least one choke. For high – pressure wells it is desirable to have a positive choke in series with an adjustable choke. On offshore facilities and other high – risk situations, an automatic shutdown valve should be installed on the wellhead. In all cases, block valves are needed so that maintenance can be performed on the choke if there is a long flowline. Whenever flows from two or more wells are commingled in a central facility, it is necessary to install a manifold to allow flow from any one well to be produced into any of the bulk or test production systems.

2. If the pressure for initial separation is too high, too many light components will stay in the liquid phase at the separator and be lost to the gas phase at the tank. If the pressure is too low, not as many of these light components will be stabilized into the liquid at the separator and they will be lost to the gas phase.

3. As more stages are added to the process there is less and less incremental liquid recovery. The diminishing income for adding a stage must more than offset the cost of the additional separator, piping, controls, space, and compressor complexities. It is clear that for each facility there is an optimum number of stages. In most cases, the optimum number of stages is very difficult to determine as it may be different from well to well and it may change as the wells' flowing pressure declines with time.

4. The choice depends on the expected flowing characteristics of the wells. If large amounts of water are expected with the high – pressure wells, it is possible that the size of the other separators could be reduced if the high – pressure separator was three – phase.

5. At onshore locations the oil may be treated in a big "gunbarrel" (or settling) tank. All tanks should have a pressure/vacuum valve with flame arrester and gas blanket to keep a positive pressure on the system and exclude oxygen. This helps to prevent corrosion, eliminate a potential safety hazard, and conserve some of the hydrocarbon vapors.

Ⅱ. 1. C 2. B 3. A 4. D 5. A 6. D 7. B 8. D 9. C 10. A

Ⅲ. 1. B 2. D 3. C 4. C 5. A 6. C 7. C 8. D 9. B 10. B

Ⅳ. 1. 油嘴的开孔直径决定了油流流量,因为油流的上游压力主要取决于油井的自喷油压,而下游压力则主要取决于采油过程中第一个分离器上的压力控制阀的压力大小。

2. 若可调油嘴失控,则固定油嘴可取而代之,并将采出的油流量控制在规定范围内。

3. 只要油流是从两口以上的油井混合流入一个处理中心的,都有必要安装管汇,以便使任一油井所采油流能进入任一容器或其他采油测试系统中去。

4. 由于采出的流体具有多组分特性,因此由闪蒸计算过程得知:初级分离后,分离次数越多,越多的轻组分就会稳定在液相中。

5. 通过配置管汇及初级分离器的不同工作压力,不仅有利于高压井液的多级分离,也会保存油层能量。

V.1. 原油处理过程中分离级数增加得越多,所能回收的液体量就越少。因分离级数增加而相应减少的回收量应能抵得上安装分离器和管路的成本、占用空地的成本及压缩机组的成本。很明显,每一设备均有最佳分离级数。大多数情况下,由于油井和油井的情况不同,而且同一口油井的油流压力也会随时间的推移而下降,因此,要确定最佳分离级数是很困难的。

2. 一般情况下,由于自喷油压不同,故采用多级分离装置。自喷油压的大小可能因油井所在的油层条件不同而各异,或者即使在同一油层但产水量不同。通过配置管汇及初级分离器的不同工作压力,不仅有利于高压井液的多级分离,也会保存油层能量。高压井可将油气输至用户而无需加压;而低压自喷井可将油气输至需要少量加压的系统中。

1.3　两相油气分离

🔲 导语

分离设备是设计用来把采出流体分离成油、气和水的容器,主要有分离器、捕集器、分离罐、闪蒸容器、洗涤器和过滤器。分离设备通常用于油井或油罐区附近的矿场或生产平台,将油气井采出流体分离为油和气或者液体和气体。本文主要介绍两相分离器,以及达到理想分离效果的设计要求,还将介绍多种机械设备是如何利用采出流体的物理作用来达到理想分离效果的。

🔲 课文

从井口采出的流体是由密度、蒸气压以及其他物理特性不相同的碳氢化合物组成的复杂混合物。当井下流体从高温、高压的油层中流出时,温度和压力都会降低。天然气从液体中逸出,此时出井流体在性质上发生变化:气体中携带液体小滴,而液体中含有气泡。在油气开采、加工、处理过程中,对各相流体进行物理分离是一项主要作业内容。

在油气分离作业的设计中,我们可以采用机械方法,从碳氢化合物的流体中将只能在特定温度和压力下存在的原油及天然气组分分离出来。由于分离用容器在任何采油厂通常都是初级的处理容器,故合理地进行分离器设计很重要;如果这些油气处理设备设计得不合理,就会成为全部采油作业中的瓶颈,并会使采油厂减产。

若所用分离器只是将天然气从总的液流中分离出来,这种分离器叫两相分离器;若所用分离器除将天然气分离出来外,还将液相中的原油和水分离开来,这种分离器叫三相分离器。当气液比很高时,有时称分离器为气体洗涤器。某些操作人员称直接处理出油流的分离器为"捕集器"。但无论如何,分离器均有相同的结构,并且根据相同的设计方法来确定尺寸。

卧式分离器

分离器可设计为卧式、立式及球形三种结构。图1.3.1为卧式分离器示意图。流体进入分离器并且冲在入口分流器上,使其动能突然改变。液体及其中气体在入口分流器上得到初次分离。重力作用使液体小滴从气体中脱出来,落至分离器的底部。液体聚集段使带进来的天然气有足够的保留时间从原油中逸出,并上升至蒸气段。此外,该分离器也提供了缓冲容积,若有必要,可用以处理液流中的间歇性水击现象。然后,该液流通过液体放空阀从容器中流出去。该液体放空阀可由液面控制器调节。后者随着液面的变化,对放空阀进行相应控制。

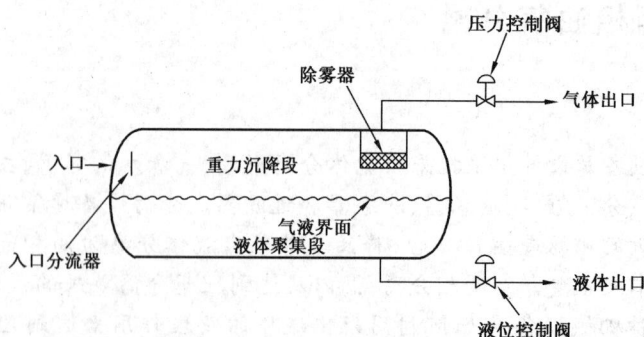

图 1.3.1　卧式分离器示意图

天然气经过入口分流器,然后横向穿过液体上方的重力沉降段。当天然气进入并通过此段时,天然气中携带的虽然经过入口分流器但仍未全部分离出来的小雾滴就会靠重力而落至气液界面处。

某些直径更小的雾滴,在穿过重力沉降段时也不易分离出来。天然气离开分离器之前,会经过聚结段或除雾器。该段利用叶片、金属丝网或金属板等零件,在气体离开分离器之前,将液体小雾滴聚结或吸去进行最后的分离。

分离器中的压力靠压力控制器维持。压力控制器对分离器中的压力变化起传感作用,并且分别发出信号以控制压力阀的开或关。通过控制天然气在分离器中蒸气段的流速,来维持分离器压力。通常,卧式分离器工作时液体占一半,以保证气液界面的表面积为最大。

立式分离器

图1.3.2是立式分离器的示意图。在此结构中,流体从分离器的侧面进入。和卧式分离器的结构一样,入口分流器对气液进行初次分离。液体向下流至分离器的底部即液体聚集段。液体连续向下流,通过聚集段向出口流走。当气液达到平衡时,气泡流向相反方向,并且最后上升进入蒸气段。液位控制器及液体放空阀的操作和卧式分离器中的相同。

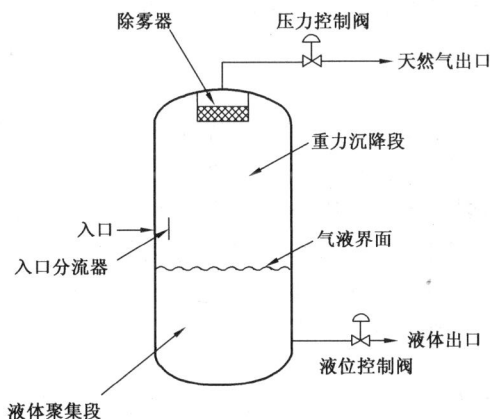

图1.3.2 立式分离器示意图

天然气经过入口分流器后垂直向上流向出口。液滴在重力沉降段与天然气的走向相反,垂直向下。天然气离开分离器之前,还要经过除雾器。压力及液位的控制方法与卧式分离器相同。

球形分离器

常规球形分离器见图1.3.3。在此类型分离器中,也有上述4个工作段。球形分离器可看作没有中间圆筒壳而只有上下两个球形帽的立式分离器特例。从承压角度来说,球形分离器是较为理想的,但存在的问题是:(1)防止液体水击现象的能力较差;(2)制造困难,油田很少采用。由于上述原因,就不对球形分离器进行深入介绍了。

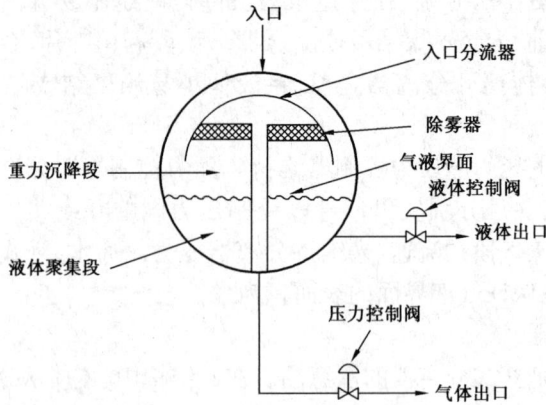

图 1.3.3　球形分离器示意图

其他结构的分离器

旋转分离器在设计上是利用离心力来工作的,适用于非常洁净的气流。气流进入分离器,通过离心力旋转着把液体和尘土分离出去。虽然这种结构的分离器尺寸较小,但在采油作业中很少使用,这是因为:(1)这种分离器的设计对流量非常敏感;(2)这种分离器比前述常规分离器需要较大的压降。因为速度降低,分离效果就会降低,旋转分离器不适合于流量变化范围较大的油田。这些装置通常用来回收脱水塔中携带乙二醇的下游流体。近年来,由于空间和重量上的优势,使用旋转式分离器的需求增大。

处理小流量流体时,通常采用双筒式分离器。双筒式分离器的气体筒和液体筒是分开的。另有一种适用于高气流、低液流的分离器,叫过滤分离器,有卧式和立式两种结构。

⟋ Key to Exercises

Ⅰ.1. Proper separator design is important because a separation vessel is normally the initial processing vessel in any facility, and improper design of this process component can"bottleneck"and reduce the capacity of the entire facility.

2. Separators are classified as "two – phase" if they separate gas the total liquid stream and "three – phase" if they also separate the liquid stream into its crude oil and water components.

3. Separators are designed in either horizontal, vertical, or spherical configurations.

4. The pressure in the separator is maintained by a pressure controller. The pressure controller senses changes in the pressure in the separator and sends a

signal to either open or close the pressure control valve accordingly. By controlling the rate at which gas leaves the vapor space of the vessel the pressure in the vessel is maintained. Normally, horizontal separators are operated half full of liquid to maximize the surface area of the gas liquid interface.

5. Because (1) their design is rather sensitive to flow rate and (2) they require greater pressure drop than the standard configurations previously described. Since separation efficiency decreases as velocity decreases, cyclone separators are not suitable for widely varying flow rates. These units are commonly used to recover glycol carryover downstream of a dehydration tower. In recent years, demand for using cyclone separators on floating facilities has increased because space and weight considerations are overriding on such facilities.

Ⅱ. 1. B 2. C 3. A 4. B 5. D 6. C 7. A 8. D 9. C 10. B

Ⅲ. 1. A 2. A 3. D 4. B 5. B 6. C 7. C 8. D 9. D 10. A

Ⅳ. 1. 从井口采出的流体是由密度、蒸气压以及其他物理特性不相同的碳氢化合物组成的复杂混合物。

2. 由于分离用容器在任何采油厂通常都是初级的处理容器,故合理地进行分离器设计很重要;如果这些油气处理设备设计得不合理,就会成为全部采油作业中的瓶颈,并会使采油厂减产。

3. 通过控制天然气在分离器中的蒸气段的流速,来维持分离器压力。

4. 气流进入分离器,通过离心力旋转着把液体和尘土分离出去。

5. 虽然这种结构的分离器尺寸较小,但在采油作业中很少使用,这是因为:(1)这种分离器的设计对流量非常敏感;(2)这种分离器比前述常规分离器需要较大的压降。因为速度降低,分离效果就会降低,旋转分离器不适合于流量变化范围较大的油田。

Ⅴ. 1. 本文主要介绍两相分离器,以及达到理想分离效果的设计要求,还将介绍多种机械设备是如何利用采出流体的物理作用来达到理想分离效果的。

2. 若所用分离器只是将天然气从总的液流中分离出来,这种分离器叫两相分离器;若所用分离器除将天然气分离出来外,还将液相中的原油和水分离开来,这种分离器叫三相分离器。

3. 和卧式分离器的结构一样,入口分流器对气液进行初次分离。液体向下流至分离器的底部即液体聚集段。液体连续向下流,通过聚集段向出口流走。当气液达到平衡时,气泡流向相反方向,并且最后上升进入蒸气段。液位控制器及液体放空阀的操作和卧式分离器中的相同。

1.4 油气水三相分离

导语

三相分离器的工作原理是要分离的三种组分因密度不同缓慢运动时会自然分层,天然气会逐渐位于最上面,水位于底部而油处于中间。任何沙子类的固体都会沉到分离器底部。三相分离器可设计成卧式或立式压力容器。由于进入三相分离器的流体或者直接来自生产井,或者来自高压分离器,所以设计容器时不仅要考虑油与水的分离,而且要考虑液体中溶解气的闪蒸分离。

课文

气液两相分离的设计理念也适用于油气水三相分离。当油与水混合达一定程度并进行沉降后,容器底部会出现一层比较清洁的游离水。这层游离水厚度按一定曲线随时间增加,经过一段时间后,比如约 3~20min,此水层厚度再也没有显著变化。靠重力沉降分离出的这部分水称为"游离水"。正常情况下,先分离出游离水,再设法处理残余油和乳状液层是比较有利的。"三相分离器"及"游离水脱除器"两个术语指的是设计用来从原油和水的混合物中分离出游离水的压力容器。

卧式分离器

三相分离器可设计成卧式或立式压力容器。图 1.4.1 是卧式三相分离器示意图。油气水混合物进入分离器,撞击入口折流板,动量的突然变化使液体和气体发生初步分离。在大多数设计中,入口折流板附有一根降液管,使液体流到油气界面以下。

分离器集液部分使油和乳状液有充足的时间在顶面形成一层"油层",游离水沉降到容器底部。图 1.4.1 是装有界面控制器和隔油挡板的典型卧式分离器,挡板用于维持油面,界面控制器用于维持水面,高过挡板的油被撇去,挡板下游的油面靠操纵出油阀的液面控制器控制。

水从位于隔油挡板上游的短管排出。界面控制器检测出油水界面高度,再将信号传送给排水阀,允许适量的水排出分离器,使油水界面维持在设计高度。

气体在分离器内水平流动,流经除雾器和压力控制阀后排出容器。压力控制阀使容器维持恒定的压力。根据气液分离的具体要求,油气界面高度可控制在容器直径的 1/2 至 3/4 范围内,常用结构的液面高度为容器直径的一半。

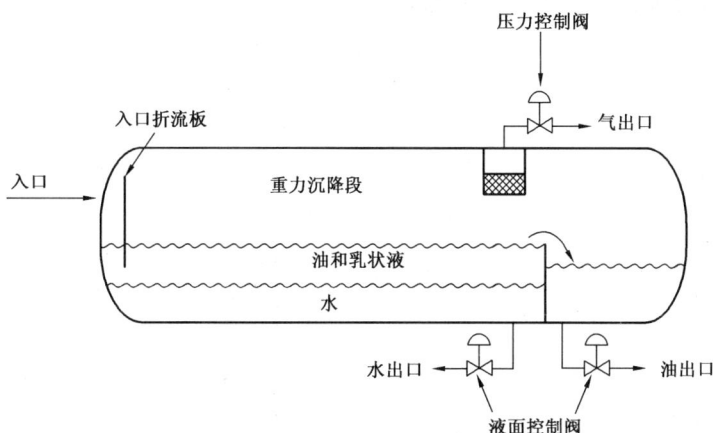

图 1.4.1　卧式三相分离器示意图

图 1.4.2 是另一种称为"油槽挡板式"结构的设计,按此设计不必安装液体界面控制器,油和水都在挡板上流过。在挡板处,液面控制是由一个简单的平衡浮子来完成的。油面高出隔油挡板时,油溢出流入油槽,油槽内油面由操纵出油阀的液面控制器控制。水在油槽下面流动,再流过挡水板,挡水板下游液面由操纵排水阀的液面控制器控制。

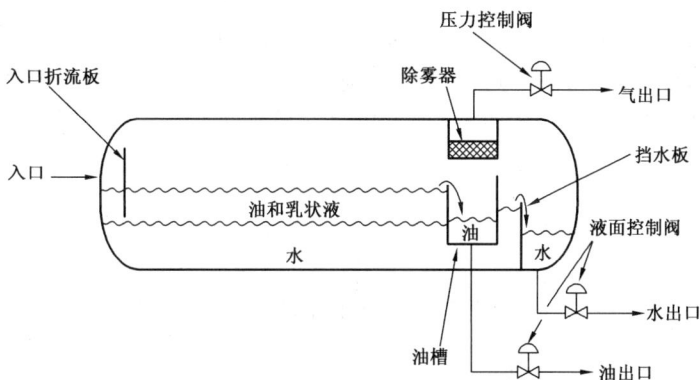

图 1.4.2　油槽挡板设计

隔油挡板高度控制了容器中的液面。因为油和水的相对密度差,隔油挡板与挡水板的高度差控制了油层厚度。分离器操作中十分关键的问题是挡水板高度必须充分低于隔油挡板高度,这样油层厚度才能为原油在分离器内提供充足的停留时间。假如挡水板太低,且油水相对密度差小于预期,那么,油层厚度增加过大,就会致使油从油槽下面流过,并与水一起从排水口排出。正常情况下,隔油挡板或挡水板都做成高度可调的,在油水相对密度发生改变或者流量变化时可进行调节。

界面控制的优点在于油水相对密度或流量发生意外变化时容易进行调节。然而,在重质原油或可能有大量乳状液或石蜡的情况下,可能难以检测出界面。此时,推荐使用油槽挡板控制。

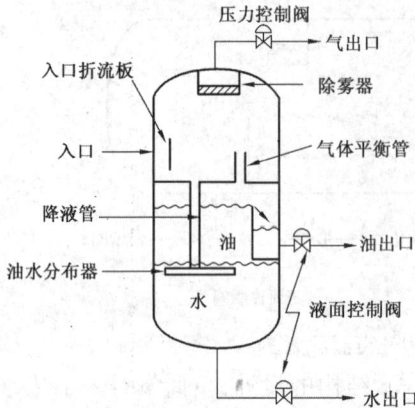

图1.4.3 立式三相分离器示意图

立式分离器

图1.4.3是典型的立式三相分离器结构。流体从侧面流入容器,与卧式分离器一样,入口折流板使大部分气体分离出来,需要一根降液管引导液体向下流动穿过油气界面,而不致影响中间部位的撇油、集油。另外,还需要一根气体平衡管来均衡容器下部与上部气室两者之间的气体压力。

油水分布器或降液管下端出口位于油水界面处。以此高度为界,油滴上升时,油相内束缚的游离水都将分离出来。水滴与油滴逆向流动。同样,水滴下沉时,水相中束缚的油滴往上浮。

图1.4.4是立式分离器上常用的三种不同的控制方法。第一种方法是严格控制液面。用普通平衡浮子控制油气界面,调节集油段放油控制阀;用界面浮子控制油水界面,调节排水控制阀。由于容器内部没有安装折流板或挡板,此种结构最容易加工制作,并能较好地处理沙子和固体杂质比较多的流体。

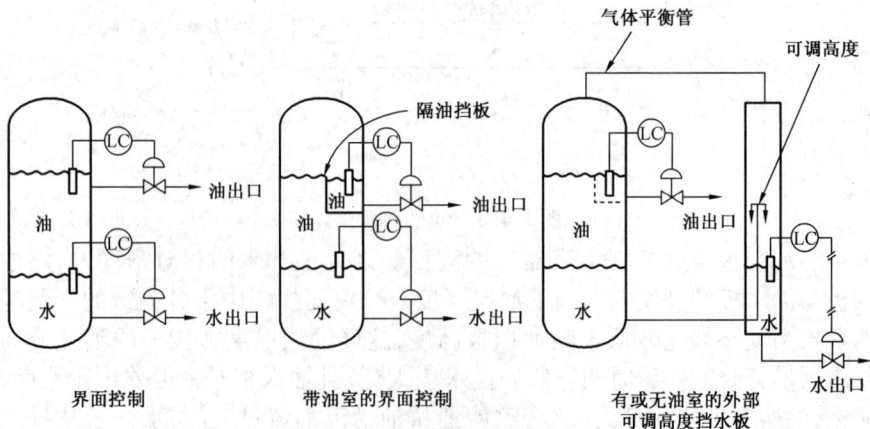

图1.4.4 液油控制方案

　　第二种方法是用隔油板在固定位置控制油气界面。由于油滴必须上升到挡板高度才能排出容器,所以油水能较好地分离。其缺点是油室占据了容器的有效容积,并增加了加工制作费用。此外,沉积物和固体杂质容易聚集在油室里难以排出,可能需要单独加装一个低液面停机装置以防出油阀无法开启。

　　第三种方法是用两块挡板,这样就不需要界面浮子了。界面位置是由外部挡水板高度相对于隔油板或出油口的高度控制的。这与卧式分离器的油槽挡板设计相似。此种结构的优点是取消了界面控制,缺点是需要增加外部管路和空间。

卧式与立式分离器的选择

　　如上所述,每种设计都有各自的优缺点。与两相分离一样,在三相分离中,从工艺角度出发,卧式分离器中的流动几何形态确实可能有利些。然而,在特定应用场合,之所以选用立式分离器,可能并非出于工艺上的考虑。

Key to Exercises

　　Ⅰ.1. Because flow normally enters these vessels either directly from（1）a producing well or（2）a separator operating at a higher pressure, the vessel must be designed to separate the gas that flashes from the liquid as well as separate the oil and water.

　　2. Interface control has the advantage of being easily adjustable to handle unexpected changes in oil or water specific gravity or flow rates. However, in heavy oil applications or where large amounts of emulsion or paraffin are anticipated, it may be difficult to sense interface level. In such a case bucket and weir control is recommended.

　　3. This design eliminates the need for a liquid interface controller. Both the oil and water flow over weirs where level control is accomplished by a simple displacer float. The oil overflows the oil weir into an oil bucket where its level is controlled by a level controller that operates the oil dump valve. The water flows under the oil bucket and then over a water weir. The level downstream of this weir is controlled by a level controller that operates the water dump valve.

　　4. Interface control has the advantage of being easily adjustable to handle unexpected changes in oil or water specific gravity or flow rates. However, in heavy oil applications or where large amounts of emulsion or paraffin are anticipated, it may be difficult to sense interface level. In such a case bucket and weir

control is recommended.

5. The first method is to use a regular displacer float to control the gas – oil interface and regulate a control valve dumping oil from the oil section. The second method is to use a weir to control the gas – oil interface level at a constant position. The third method uses two weirs, which eliminates the need for an interface float.

Ⅱ. 1. A 2. D 3. B 4. C 5. D 6. C 7. B 8. A 9. C 10. D

Ⅲ. 1. D 2. A 3. C 4. A 5. B 6. C 7. D 8. C 9. C 10. D

Ⅳ. 1. 气液两相分离的设计理念也适用于油气水三相分离。

2. "三相分离器"及"游离水脱除器"两个术语指的是设计用来从原油和水的混合物中分离出游离水的压力容器。

3. 界面控制器再将信号传送给排水阀,允许适量的水排出分离器,使油水界面维持在设计高度。

4. 由于油滴必须上升到挡板高度才能排出容器,所以油水能较好地分离。

5. 与两相分离一样,在三相分离中,从工艺角度出发,卧式分离器中的流动几何形态确实可能有利些。

Ⅴ. 1. 当油与水混合达一定程度并进行沉降后,容器底部会出现一层比较清洁的游离水。这层游离水厚度按一定曲线随时间增加,经过一段时间后,比如约3~20min,此水层厚度再也没有显著变化。靠重力沉降分离出的这部分水称为"游离水"。正常情况下,先分离出游离水,再设法处理残余油和乳状液层是比较有利的。

2. 隔油挡板高度控制了容器中的液面。因为油和水的相对密度差,隔油挡板与挡水板的高度差控制了油层厚度。分离器操作中十分关键的问题是挡水板高度必须充分低于隔油挡板高度,这样油层厚度才能为原油在分离器内提供充足的停留时间。假如挡水板太低且油水相对密度差小于预期,那么,油层厚度增加过大,致使油从油槽下面流过,并与水一起从排水口排出。

第2章 原油及采出水处理

2.1 乳状液处理

⊡ 导语

正常的油田乳状液以油为连续相即外相,水为分散相即内相。在一些高含水的个别情况下,有可能形成水为连续相而油滴为内相的反相乳状液。据报道,低密度原油和稠油中有比较复杂的乳状液,这些复杂的乳状液不仅含有水外相,而且在分散油滴里还包含水内相。由于大多数原油处理系统处理的是正常的油包水乳状液,所以本文主要讨论这种处理系统。

⊡ 课文

乳状液的形成

乳状液的形成条件包括两种互不相溶的液体和一种乳化剂,并要充分搅动,使非连续相分散到连续相中。石油开采中,油与水是两种互不相溶的液体,地层流体中几乎总是含有小固体颗粒、石蜡、沥青质等乳化剂,并且,在地层流体流入井筒、沿油管升到地面和通过地面油嘴的过程中,总会受到充分搅动。

搅动程度和乳化剂的特性与数量决定了乳状液的"稳定性"。某些稳定的乳状液如果存放在罐里不做任何处理,可能需要几周乃至几个月才能分层。而不稳定的乳状液可能在几分钟内就会分离成比较纯静的油相与水相。

油和水可以形成油包水型乳状液,其中水是分散相,油为外相;相反,它们也可以形成水包油型乳状液,其中油是分散相,水是分散介质。

乳化剂

为加深对乳状液稳定性的理解,应当说明:在纯油与纯水的混合液中,假如没有乳化剂,并且一点也不搅动,则无法形成乳状液。如果将纯油与纯水混合后倒入一个容器里,油水会很快分层的。自然状态是让不混相液体接触面最小或者表面积最小。分散在油里的水滴成球形液滴。小液滴会聚结成大液滴,体积一定时,形成的界面面积较小。如果没有任何一种乳化剂,这些水滴终将沉入器底,形成最小的界面面积。这类混合液体纯属"分散体"。

乳化剂具有表面活性。乳化剂中的某些分子特别亲油,而其他分子更易被水吸引。乳化剂往往不溶于某种液相,所以聚集于界面处。乳化剂可通过多种途径使分散体转变为乳状液。乳化剂的作用表现在以下几方面:

(1)降低水滴的界面张力,使之形成较小液滴。液滴越小,聚结成较大液滴所需要的时间越长,而且只有较大液滴才能迅速沉降。

(2)乳化剂在这些液滴表面形成一层黏性包膜,在小液滴碰撞时,这层包膜阻止小液滴聚结成较大液滴。由于聚结过程受阻,搅动形成的小液滴需要更长时间才能沉降下来。

(3)乳化剂可能是以某种方式排列的极性分子,使液滴表面带电,同性相斥,要是两滴液滴聚结起来,必须有足够大的力克服这种斥力。

通常,原油中有许多天然的表面活性物质可以起到乳化剂的作用。石蜡、胶质、有机酸、金属盐类、胶肽粉砂和黏土、沥青质都是油田常见的乳化剂。修井液和钻井液也是乳化剂的来源。

乳化剂的类型和数量对乳状液的稳定性有直接影响。乳状液温度变化过程也很重要,因为它影响石蜡和沥青质的形成。乳化剂运移到油水界面的速度和界面结合力强度也是重要因素。在搅拌或产生石蜡和沥青质后立即形成的乳状液稳定性较差,加入乳化剂不能运移到界面,此时乳状液是比较容易处理的。老化的乳状液可能变得更难处理。通常,原油黏度越低,相对密度越小,老化过程就越快。因此,在处理低黏度、高 API 度原油时,早期处理作用不大。

破乳剂

用于处理乳状液的高效化学破乳剂有各种各样的商标,如 Tretolite™、Visco™ 和 Breaxit™。破乳剂可起到抵消乳化剂的作用,但它们是表面活性剂,因此,用量过度会降低水滴的表面张力,实际上会形成更稳定的乳状液。

破乳剂要起到 4 个方面的重要作用:

(1)强烈地吸引油水界面。

(2)絮凝。

(3)聚结。

(4)固体润湿。

同时具备这些作用时,破乳剂就会促进油与水的分离。破乳剂必须有能力迅速穿过油相运移到液滴界面,在此,破乳剂必须与浓度更大的乳化剂抗争。破乳剂对具有相似条件的液滴必须也有吸引力。大团液滴以此方式聚集到一起,在显微镜下,形如成团的鱼子。油呈现光亮状,因为使光线散射的小液滴已不复存在,此时,乳化剂膜依然呈连续状。假如乳化剂很弱,絮凝力就足以引起小液滴聚结,但在大多数情况下并非如此。所以,破乳剂必须抵消乳化剂的作用,并促进液滴界面膜破裂,这正是引起小液滴聚结的起始条件。当乳状液处于絮凝状态下,这层膜破裂,导致水滴直径迅速增大。

破乳剂抵消乳化剂的过程取决于乳化剂的类型。硫化铁、黏土、钻井液都是亲水的,亲水性使它们离开界面并扩散到水滴里。石蜡与沥青质可以溶解或者发生变化,使其膜黏度降低。这样发生碰撞时,它们可以向旁边流出,或变为亲油,分散在油里。

通常,要使一种化学结构产生以上所有的4种理想作用是很难的,因此,用几种药剂的混合配方可综合发挥各自的有效作用。

破乳剂的选择应依工艺系统而异。假如处理工艺用沉降罐,选用比较长效的药剂可取得良好结果;另一方面,假如系统采用电化学处理工艺,靠电场作用完成部分絮凝和聚结过程,就需要选用速效剂而不是具有液滴增大作用的药剂。

现场条件改变时,化学药剂的需求也应改变。假如工艺条件发生变化,例如使用静电脱水器且流量非常小时,化学药剂的需求也应改变。四季气温的变化会引起石蜡诱发的乳状液问题。修井作业增加了油中的固体杂质含量,改变了乳状液的稳定性。因此,不管一种破乳剂在某段时间里多么令人满意,也无法设想其在油田开采的全过程中总是有效的。

🔲 Key to Exercises

Ⅰ.1. Normal emulsions are complex emulsions in low gravity, viscous crudes that contain a water external phase and have an internal water phase in the dispersed oil.

2. The action of the emulsifier can be visualized as one or more of the following:

(1) It decreases the interfacial tension of the water droplet, thus causing smaller droplets to form. The smaller droplets take longer to coalesce into larger droplets, which can settle quickly.

(2) It forms a viscous coating on the droplets that keeps them from coalescing into larger droplets when they collide. Since coalescence is prevented, it takes longer for the small droplets created by agitation to settle out.

(3) The emulsifiers may be polar molecules, which align themselves in

such a manner as to cause an electrical charge on the surface of the droplets. Since like electrical charges repel, two droplets must collide with sufficient force to overcome this repulsion before coalescence can occur.

3. Naturally – occurring surface active materials normally found in crude oil serve as emulsifiers. Paraffins, resins, organic acids, metallic salts, colloidal silts and clay, and asphaltenes are common emulsifiers in oil fields. Workover fluids and drilling mud are also sources of emulsifying agents.

4. They are strong attraction to the oil – water interface, flocculation, coalescence and solid wetting.

5. The demulsifier selection should be made with the process system in mind. If the treating process is a settling tank, a relatively slow – acting compound can be applied with good results. On the other hand, if the system is a chemelectric process where some of the flocculation and coalescing action is accomplished by an electric field, there is need for a quick – acting compound, but not one that must complete the dropletbuilding action.

Ⅱ. 1. D 2. A 3. C 4. B 5. B 6. D 7. C 8. B 9. A 10. D

Ⅲ. 1. C 2. B 3. D 4. D 5. C 6. A 7. C 8. D 9. B 10. D

Ⅳ. 1. 乳状液的形成条件包括两种互不相溶的液体和一种乳化剂,并要充分搅动,使非连续相分散到连续相中。

2. 乳化剂可能是以某种方式排列的极性分子,使液滴表面带电。

3. 破乳剂通常是表面活性剂,因此,用量过度会降低水滴的表面张力,实际上会形成更稳定的乳状液。

4. 石蜡与沥青质可以溶解或者发生变化,使其膜黏度降低。这样发生碰撞时,它们可以向旁边流出,或变为亲油,分散在油里。

5. 另一方面,假如系统采用电化学处理工艺,靠电场作用完成部分絮凝和聚结过程,就需要选用速效剂而不是具有液滴增大作用的药剂。

Ⅴ. 1. 破乳剂抵消乳化剂的过程取决于乳化剂的类型。硫化铁、黏土、钻井液都是亲水的,亲水性使它们离开界面并扩散到水滴里。石蜡与沥青质可以溶解或者发生变化,使其膜黏度降低。这样发生碰撞时,它们可以向旁边流出,或变为亲油,分散在油里。

2. 乳化剂的类型和数量对乳状液的稳定性有直接影响。乳状液温度变化过程也很重要,因为它影响石蜡和沥青质的形成。乳化剂运移到油水界面的速度和界面结合力强度也是重要因素。在搅拌或产生石蜡和沥青质后立即形成的乳状液稳定性较差,加入乳化剂不能运移到界面,此时乳状液是比较容易处理的。老化的乳状液可能变得更难处理。通常,原油黏度越低,相对密度越小,老化过程就越快。因此,在处理低黏度、高 API 度原油时,早期处理作用不大。

2.2　原油处理系统

导语

原油脱水不能单纯依靠重力分离,常常需要进行额外处理。通常,用加热方法分离"油包水"乳状液。聚结过程要求水滴有充分的时间互相接触,聚结起来的水滴上的重力可能足以使这些水滴沉降到处理器底部。因此,设计时有必要考虑温度、时间、抑制水滴沉降的原油黏滞特性以及决定水滴沉降速度的容器实际尺寸等。

课文

为使原油达到外输要求,必须确定处理原油最理想的方法。为此,在选择处理系统时,应考虑若干因素,如:

(1)乳状液的稳定度;

(2)原油与采出水的相对密度;

(3)原油、采出水、井口天然气的腐蚀性;

(4)采出水的结垢趋势;

(5)要处理的流体量及其含水率;

(6)原油的结蜡趋势;

(7)设备的理想操作压力;

(8)采出的井口天然气售价及是否有外销系统。

通常,用加热方法分离"油包水"乳状液。两种不相溶液体的温度提高可削弱乳化剂的作用,使分散水滴互相碰撞,小水滴碰撞后合并,水滴直径增大并开始沉降。如果设计正确,那么,因为存在油水相对密度差,水将沉到处理器底部。

处理设备

立式处理器

最常用的单井矿场处理器是立式处理器,如图 2.2.1 所示。流体从处理器顶部流进气体分离段。设计气体分离段尺寸时必须格外小心,使流入的流体有足够空间分离出气体。假如油气分离器在处理器下部,气体段尺寸可能非常小。气体分离段应装有入口分流器和除雾器。

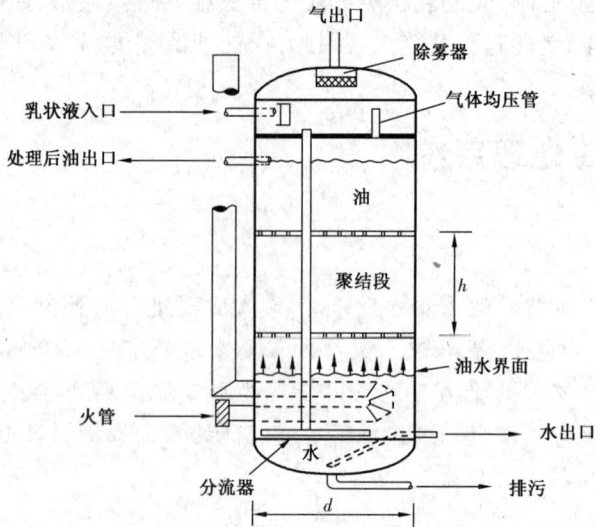

图 2.2.1 立式处理器示意图

　　流体通过降液管流到处理器底部,即游离水脱除段。如果游离水脱除段在处理器的下部,底部空间就会很小。如果井流需要全部处理,那么设计游离水脱除段尺寸时应该使油和水都有 3～5min 的停留时间,使游离水沉降出来,这样就可以将加热通过加热段上升的液流所需的燃料气量减到最少。降液管的末端应稍低于油水界面,使正在处理的油受到"水洗"。这将有助于油中水滴的聚结。

　　油和乳状液上升到加热火管上方,进入聚结段,在此停留的时间足以使油连续相中的小水滴聚结起来,沉降到底部。

　　处理后的原油经出油口排出,由于加热而从油里闪蒸出的气体经气体平衡管进入气室。油面靠气动装置或机械杠杆操纵的排泄阀控制,油水界面由界面控制器或外部安装的可调高度出水管控制。

　　设备供应商应负责处理器的详细设计,包括各种内部构件的设计。图 2.2.2 是油水分离罐,是常压罐中的纵向流动处理器。通常,油水分离罐顶部是气室,气液在此分离,气体放空,罐内还有一根降液管。因为油水分离罐的直径一般比立式加热脱水器的直径大得多,所以罐内大多装有精心设计的油水分流系统,使乳状液均匀地向上流动,这样可充分利用整个横截面。大多数油水分离罐是不加热的,不过也可以在罐外预热进罐流体,或在罐内加装加热盘管,或设计一个加热回路使热水闭路循环。最好采取第一种方法,既可以使入口物流中的游离水加热沉降,又有利于气体释放出来。

　　以前,陆上低产老油田常用油水分离罐。近年来,立式加热脱水器造价

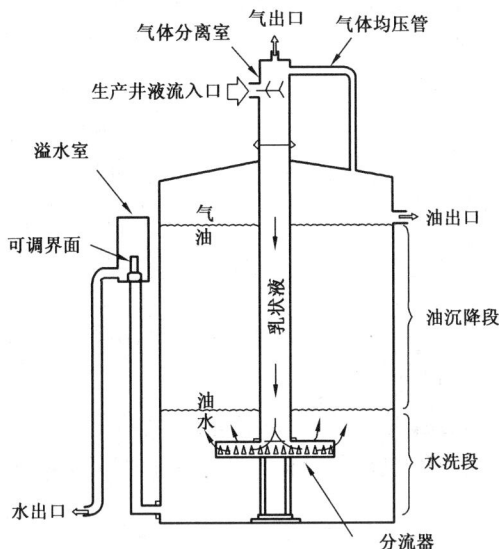

图 2.2.2　油水分离沉降罐

降低,取代了单井用油水分离罐。气候温暖地区的大型陆上装置仍普遍使用油水分离罐,在冬季寒冷地区也需要通过加热使大量原油高于倾点,以防事故发生,不过使用加热式油水分离罐成本太高。

卧式处理器

多井集中处理站大多需要用卧式处理器。图 2.2.3 是典型的卧式处理器设计图。

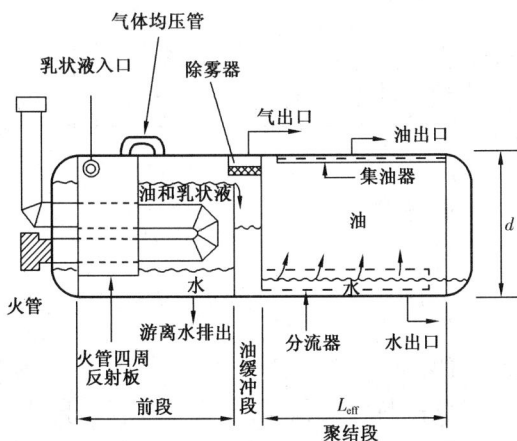

图 2.2.3　卧式加热脱水处理器

流体进入处理器前室,气体在此闪蒸。液体滴入油水界面附近,液体在此受到"水洗",分离出游离水。油和乳状液上升流经火管,被撇入油缓冲室。处理器入口室的油水界面由界面控制器控制,控制器操纵游离水泄放阀。

油和乳状液通过油水分流器流入后室,即处理器的液封聚结段。分流器使流体在此段全部均匀分布。处理后的油通过集油装置集中在顶部,该装置大小足以使油均匀地纵向流动。正在聚结的水滴自上而下沉降,与正在上升的油连续相逆向而行。油水界面由该段专用液面控制器和排泄阀保持。

油缓冲室里的液面控制装置操纵出油管路中的排泄阀,调节处理器顶部排出油量,维持液封条件。

入口室大小应满足游离水沉降和原油加热的需要。应认真确定聚结段的大小,以便为聚结提供充足的停留时间,使正在聚结的水滴能与上升油流逆向沉降。

静电处理器

有些处理器有电极段。图2.2.4是一台卧式静电处理器的典型示意图。流体在静电处理器中的流动路线与卧式处理器相同,唯一的区别在于用交流或直流静电场来促进水滴聚结。

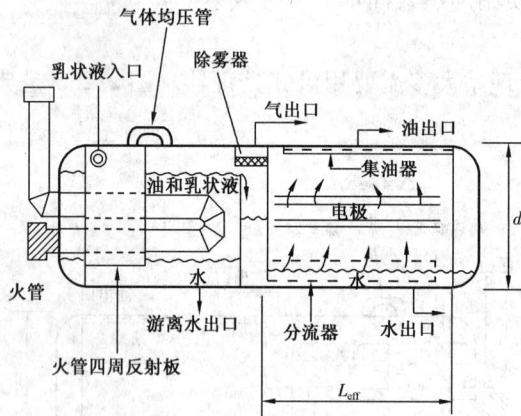

图2.2.4 卧式静电处理器

静电聚结处理器的设计步骤未公开发布。液滴在电场中的聚结过程依赖于所处理的特定乳状液特性,所以要推算出用于沉降方程中水滴直径的通用关系式是不太可能的。现场应用经验表明,静电处理器可有效地降低原油含水量,使原油中沉淀物和水的含量低于0.5%~1.0%,因此在脱盐应用中特别有吸引力。然而,正常原油处理中,0.5%~1.0%沉淀物和水的含量是可以接受的,推荐选用加热脱水器的尺寸。安装后逐步进行调试,使电极板可在较低温度下有效地处理乳状液。

Key to Exercises

Ⅰ.1. They are tightness of the emulsion, specific gravity of the oil and produced water, corrosiveness of the crude oil, produced water, and casinghead gas, scaling tendencies of the produced water, quantity of fluid to be treated and percent water in the fluid, paraffin – forming tendencies of the crude oil, desirable operating pressures for equipment and availability of a sales outlet and value of the casinghead gas produced.

2. Design considerations should necessarily include temperature, time, viscous properties of oil that inhibit settling, and the physical dimensions of the vessel, which determine the velocity at which settling must occur.

3. The common method for separating the " water – in – oil" emulsion is to heat the stream. The reason is that increasing the temperature of the two immiscible liquids deactivates the emulsifying agent, allowing the dispersed water droplets to collide. As the droplets collide they grow in size and begin to settle.

4. The vertical treater is most commonly used in single – well lease while horizontal treaters are normally used for most multi – well situations.

5. The difference is that an AC and/or DC electrostatic field is used to promote coalescence of the water droplets.

Ⅱ.1. B 2. A 3. D 4. C 5. D 6. A 7. B 8. D 9. C 10. A

Ⅲ.1. C 2. C 3. A 4. B 5. B 6. D 7. A 8. C 9. B 10. D

Ⅳ.1. 为使原油达到外输要求,必须确定处理原油最理想的方法。为此,在选择处理系统时,应考虑若干因素。

2. 如果设计正确,那么,因为存在油水相对密度差,水将沉到处理器底部。

3. 降液管的末端应稍低于油水界面,使正在处理的油受到"水洗"。

4. 在冬季部分寒冷地区也需要通过加热使大量原油高于倾点,以防事故发生,不过使用加热式油水分离罐成本太高。

5. 应认真确定聚结段的大小,以便为聚结提供充足的停留时间,使正在聚结的水滴能与上升油流逆向沉降。

Ⅴ.1. 因为油水分离罐的直径一般比立式加热脱水器的直径大得多,所以罐内大多装有精心设计的油水分流系统,使乳状液均匀地向上流动,这样可充分利用整个横截面。大多数油水分离罐是不加热的,不过也可以在罐外预热进罐流体,或在罐内加装加热盘管,或设计一个加热回路使热水闭路循环。最好采取第一种方法,既可以使入口物流中的游离水加热沉降,又有利于气体释放出来。

2. 液滴在电场中的聚结过程依赖于所处理的特定乳状液特性,所以要推算出用于沉降方程中的水滴直径的通用关系式是不太可能的。现场应用经验表明,静电处理器可有效地降低原油含水量,使原油中沉淀物和水的含量低于 0.5% ~ 1.0%,因此在脱盐应用中特别有吸引力。然而,正常原油处理中,0.5% ~ 1.0% 沉淀物和水的含量是可以接受的,推荐选用加热脱水器的尺寸。

2.3 采出水处理系统(1)

🔲 导语

自 19 世纪中期,也就是最初开始钻油气井的时候,采出水处理对环境所造成的影响就有了报道。许多采出水含有高浓度的溶解离子(盐)、碳氢化合物和微量元素。这些微量元素浓度越来越高,对植物性有毒,吸附在土壤中。未经处理的污水排放可能会对周围的环境有害,导致地下水、地表水以及土壤退化,从而致使生态系统遭到破坏。本文将介绍如何选择适宜类型的采出水处理设备,以使油气作业采出水处理符合环保要求。

🔲 课文

生产操作中,常常需要处理各种各样的废水,其中包括随原油采出的污水、雨水和冲洗水。这些废水必须从原油里分离出来,并应当按照环保法规的要求处置。负责制定环保法规的组织规定了近海地区排入海里的含油污水最大含烃量,目前定为 15 ~ 50mg/L,依具体水域而异。大多数陆上油区的污水不能直接在地面排放,因为污水可能使土壤受到盐侵污染,而必须通过污水井注入适当地层或者通过地面蒸发处理,以降低水中含烃量,使之比正常的游离水脱除器或原油脱水处理器排出的污水含油量更低。

系统描述

采出水排放前总要进行某种形式的初级处理,如采用隔油罐、隔油容器、波纹板隔油池或交叉流分离器。所有这些设备都采用重力分离机理。根据处理问题的难度,可能需要二次处理,如使用波纹板隔油池或交叉流分离器或浮选设备。

　　海上含油污水经处理后可直接排出采油船外,或排入排放管桩及隔油管桩。甲板污水必须处理以除去"游离"油。陆上污水一般回注到地层里或用泵排入污水井。

　　出于安全考虑,如果设备中有排水暗管,应先接入压力容器,再进常压罐或排放管桩。为此,可采用隔油容器。系统需要除砂处理时,后两种比较理想。

处理设备

隔油罐或隔油容器

　　最简单的初级处理设备是隔油罐或隔油容器,通常设计成可为聚结和重力分离提供较长的停留时间。隔油罐或隔油容器可以是立式的,也可以是卧式的。在立式隔油罐中,油滴必须向上浮升,与向下流动的水逆行。如图2.3.1所示,有些立式隔油罐还装有入口分流器和出口集水器,使水流均匀分布。入口在油水界面以下,水里释放的少量气体有助于油滴上浮。分流器与集水器之间的滞留区会发生一些聚结,油滴的浮力使其与水流逆向浮升。油集中在水面,最终被撇去。

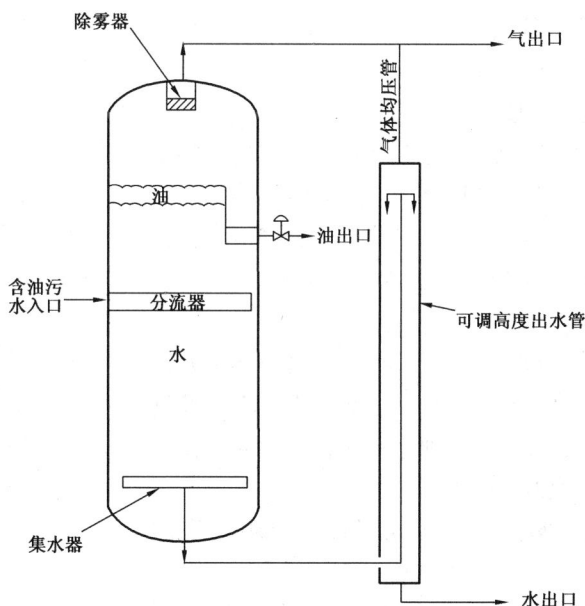

图 2.3.1　立式隔油罐示意图

　　隔油罐里的油层厚度取决于挡油板与外部安装的可调高度排水管的相对高度,还取决于两种液体的相对密度差。往往用界面控制器取代外部安装的可调高度排水管。

在卧式隔油罐中,如图 2.3.2 所示,油滴浮升方向与水流方向垂直。污水入口在油层以下,水进入后在容器中基本上呈水平流动,可用折流板校直水流方向。油滴在这段容器中发生聚结,浮升到油水界面,油被捕集,高过挡油板时最终被撇去。油面高度可用界面控制,也可用如图 2.3.1 所示的罐外安装的可调高度排水管控制,也可用油槽挡油板组控制。

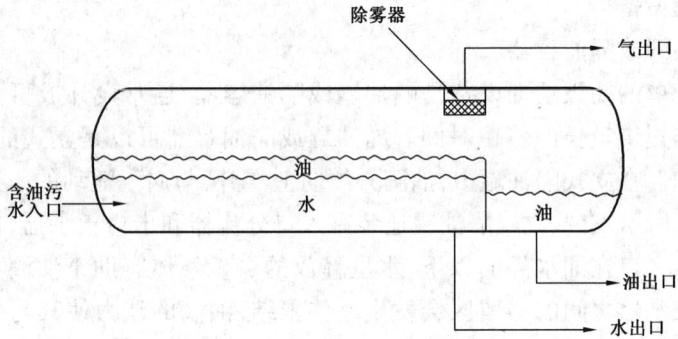

图 2.3.2　卧式隔油罐示意图

污水处理用卧式容器更有效,因为油滴与水流不用逆向流动。在下述情况下,可用立式隔油罐:

(1)必须处理砂子和其他固体颗粒,可将出水口或排砂口安排在立式容器底部,大型卧式容器上精心设计的排砂口实际使用情况并不非常满意。

(2)可能发生液体脉动。立式容器不容易因为液体脉动而误发生高液面停机。而在卧式容器中,尽管正常操作液面与高液面停机液面之间的液体体积与立式容器相等或更大些,但脉动引起的内部波浪会使液面浮子误触动停机机构。

选择压力容器还是常压隔油罐,这不仅取决于水处理要求,还要考虑系统的总要求。压力容器造价高,推荐用在以下情况:

(1)通过上游容器排泄系统可能发生液体带气现象时,会在常压放空系统产生过高的背压。

(2)水必须被泵送到更高以便进一步处理,并且安装常压容器必须用泵时。

由于使用常压容器存在潜在的超压危险和潜在的气体放空问题,应当优先选用压力容器。无论如何,必须对各自的投资和优劣做出判定。

应当提供的最短停留时间为 10 ~ 30min,以确保系统不致受到脉动的严重影响并有利于聚结。前面已讨论过,尽管延长停留时间有一定益处,但在经济上可能并不合算。隔油罐内停留时间较长时,需要用折流板改善流体分布状况,消除短路。示踪剂研究结果表明,即使装有精心设计的分流器和折流板,隔油罐内的流动形态依然很差,而且难免发生短路。这可能是因为存在密度差和温度差、固体沉积以及分流器腐蚀等。

Key to Exercises

Ⅰ. 1. In producing operations it is often necessary to handle wastewater that may include water produced with crude oil, rain water, and washdown water.

2. In offshore areas where discharge to the sea is allowed, the governing regulatory body specifies the maximum hydrocarbon content in the water that may be discharged overboard. The range is currently 15 mg/l to 50 mg/l depending on the specific location. In most onshore locations the water cannot be disposed of on the surface, due to possible salt contamination, and must be injected into an acceptable disposal formation or disposed of by evaporation.

3. The purpose of this article is to present the engineer with a procedure for selecting the appropriate type of equipment for treating oil from produced water. When this design procedure is followed, the engineer will be able to develop a process flowsheet, determine equipment sizes, and evaluate vendor proposals for any wastewater treating system.

4. Offshore, produced water can be piped directly overboard after treating, or it can be routed through a disposal pile or a skim pile. Onshore, the water will normally be reinjected in the formation or be pumped into a disposal well.

5. The pressure vessel is used instead of atmospheric skimmers in the following two situations:

(1) Potential gas blowby through the upstream vessel dump system could create too much back – pressure in an atmospheric vent system.

(2) The water must be dumped to a higher level for further treating and a pump would be needed if an atmospheric vessel were installed.

Ⅱ. 1. C 2. A 3. B 4. D 5. D 6. A 7. C 8. D 9. B 10. C

Ⅲ. 1. D 2. C 3. D 4. B 5. C 6. A 7. C 8. D 9. A 10. A

Ⅳ. 1. 负责制定环保法规的组织规定了近海地区排入海里的含油污水最大含烃量。

2. 选择压力容器还是常压隔油罐,这不仅取决于水处理要求,还要考虑系统的总要求。

3. 出于安全考虑,如果设备中有排水暗管,应先接入压力容器,再进常压罐或排放管桩。

4. 污水处理用卧式容器更有效,因为油滴与水流不用逆向流动。

5. 而在卧式容器中,尽管正常操作液面与高液面停机液面之间的液体体积与立式容器相等或更大些,但脉动引起的内部波浪会使液面浮子误触动停机机构。

Ⅴ.1. 最简单的初级处理设备是隔油罐或隔油容器,通常设计成可为聚结和重力分离提供较长的停留时间。隔油罐或隔油容器可以是立式的,也可以是卧式的。在立式隔油罐中,油滴必须向上浮升,与向下流动的水逆行。水里释放的少量气体有助于油滴上浮。分流器与集水器之间的滞留区会发生一些聚结,油滴的浮力使其与水流逆向浮升。油集中在水面,最终被撇去。

2. 尽管延长停留时间有一定益处,但在经济上可能并不合算。隔油罐内停留时间较长时,需要用折流板改善流体分布状况,消除短路。示踪剂研究结果表明,即使装有精心设计的分流器和折流板,隔油罐内的流动形态依然很差,而且难免发生短路。这可能是因为存在密度和温度差、固体沉积以及分流器腐蚀等。

2.4 采出水处理系统(2)

🔲 导语

因产层地球化学性质、碳氢化合物种类以及生产井的特性都不同,所以采出水的质量有极大差异。采出水需要处理才能提高其质量并益于使用。如果符合相应的水质标准,采出水就可以用于灌溉、畜牧、市政和工业等。按照本文提出的设计程序,工程师能绘制出工艺流程图,确定各种设备的尺寸,并对供应商提出的废水处理系统投标书作出评价。

🔲 课文

板式聚结器

板式聚结器是用内置板改善重力分离工艺的隔油罐或隔油容器。现已发明了各种构型的板式聚结器,通称为平行板隔油器、波纹板隔油器和交叉流分流器。所有这些都是根据重力分离原理使油滴浮升到板表面,在此进行聚结和油滴捕集。如图 2.4.1 所示,流体在一系列间隔很小的平行板间分流,为便于捕集油滴,平行板与水平方向成一定角度斜置。

污水排放管桩

污水排放管桩是 $24 \sim 48$in 的大口径敞口管,附装在石油平台旁,一直延伸到水下。其主要用途:(1)将平台排放的污水集中到一处;(2)污水可沿此导管排入水下较深位置,不受波浪的影响,即使在异常状况下也不会出现水面油污;(3)万一发生油漫流甲板的事故,可以报警或停机。

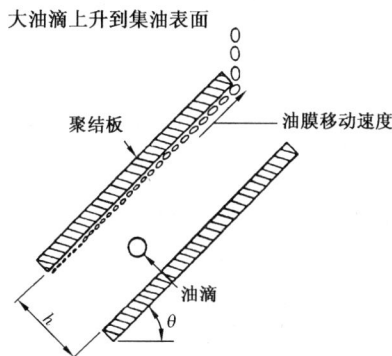

图 2.4.1 板式聚结器

环保机构要求所有采出水排入排放管桩前应经过(隔油罐、聚结器或浮选设备)处理。某些地方准许将已处理过的采出水、砂子、油滴收集盘上的液体或甲板冲洗水排入排放管桩;万一设备出现故障,排放管桩还可以作为最后一道捕集烃液的手段。

排放管桩特别适用于甲板污水排放。甲板污水来源于雨水或洗刷水,通常有油滴分散在富含氧气的淡水或盐水中。水中的氧会使其具有很强的腐蚀性,如果与采出水掺混,会在隔油罐、板式聚结器或浮选池里引起结垢或堵塞。这种污水的排放量很缺乏规律性,因此,容易导致设备出现故障。最终,这些污水在重力作用下会流到一个较低的部位集中起来,如果处理设备标高较高,应当用泵;如果处理设备标高较低,可以不用泵。为此,使用排放管桩极为理想,这样就可以防止腐蚀。通过设计,排放管桩可以安置在平台相当低的部位上,这样就可以不用污水泵,而且不会受到瞬时流量突然变化的严重影响(排放管桩可以继续工作,但排放污水的水质可能受到某种程度的影响)。排放管桩内没有任何小通道,不会发生结垢堵塞。排放管桩是污水排放的最后一道处理设备,与工艺系统不会掺混。

排放管桩用于浅水水域时长度和水深一样,在发生意外事故时可使油污容量最大化,使水面污染最小化。在深水水域,管桩应有足够的长度,以确保在管桩内充满油污前就能发出报警与停机信号。这些信号的设定位置必须有足够的高度,不致因为潮汐变化而发生误报警。

隔油管桩

隔油管桩是排放管桩中的一种。如图 2.4.2 所示,水流流经多组折流板时产生无流动区,缩短了指定油滴与主水流分离所必须上升通过的距离。油滴一旦进入此无流动区,就有充裕时间聚结和重力分离。较大油滴继续向上运移到集油挡板底侧。

隔油管桩比标准排放管桩能更有效地分离油,而且,从油分离的角度看,隔油管桩还有净化油污砂子的优点。大多数政府都立法规定,采出的砂子必须不含游离油方能排放。使用标准排放管桩时,处理容器排出的油污砂子能否符合这一规定令人怀疑。

油污砂子流经隔油管桩全长度的过程中与折流板不断摩擦并受到水的冲洗,这样可除去游离油,再在无流动的静止区捕集油滴。

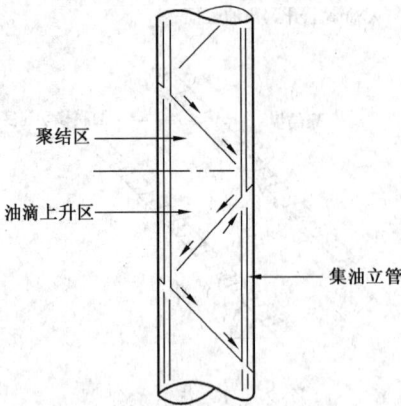

图 2.4.2　隔油管桩内的流态

（聚结区、油滴上升区、集油立管）

设备选择步骤

根据联邦政府环保条例,游离水脱除器排出的采出水至少要经过某种形式的初步处理后方可排入排放管桩或隔油管桩。甲板污水可直接排入大小相宜的排放管桩,分离掉"游离油"。

每个水处理系统的设计必须从选择尺寸开始,包括用游离水脱除器、加热脱水处理器或三相分离器进行液体分离。

除了这些限制条件外,设计工程师可以按照自己的意愿自由安排系统。上述水处理设备有多种组配方式,在某种综合条件下,可将游离水脱除器排出的含油污水直接引入隔油罐进行最终处理,然后排放。在一定情况下,可能需要一个比较完善的水处理系统,包括板式聚结器、浮选设备和隔油管桩等。方案最终评价时,主要依据设计工程师的判断能力和经验来确定设备的具体组配形式,选定尺寸。下述工作步骤只是个参考指南,不能取代工程师的判断和经验。任何情况下,都不能因为有了这些工作步骤而忽视了本领域的实际工作经验。

（1）确定采出污水的含油量。

（2）确定分散油污水的水质。

（3）确定系统采出水中油滴粒径分布状况。

（4）确定必须处理的油粒直径,以符合所要求的排放水标准。

（5）确定隔油罐尺寸。

（6）确定板式聚结器规格尺寸。

（7）根据成本和可利用的工作空间,选用隔油罐、盘管组件或板式聚结器。

(8)选择甲板污水排放方法。

Key to Exercises

Ⅰ.1. They are parallel plate interceptors (PPI), corrugated plate interceptors (CPI), or cross-flow separators.

2. Their main uses are to (1) concentrate all platform discharges into one location, (2) provide a conduit protected from wave action so that discharges can be placed deep enough to prevent sheens from occurring during upset conditions, and (3) provide an alarm or shutdown point in the event of a failure causing oil to flow overboard.

3. The disposal pile length should be as long as the water depth permits in shallow water to provide for maximum oil containment in the event of a malfunction and to minimize the potential appearance of any sheen. In deep water, the length is set to assure that an alarm and then a shutdown signal can be measured before the pile fills with oil.

4. Besides being more efficient than standard disposal piles, from an oil separation standpoint, skim piles have the added benefit of providing for some degree of sand cleaning.

5. They are : (1) Determine the oil content of the produced water influent; (2) Determine dispersed oil effluent quality; (3) Determine oil drop size distribution in the influent produced water stream; (4) Determine oil particle diameter that must be treated to meet effluent quality required; (5) Determine skimmer dimensions; (6) Determine plate coalescer dimensions; (7) Choose skim tank, SP Pack, or plate coalescer for application, considering cost and space available; (8) Choose method of handling deck drainage.

Ⅱ.1. A 2. A 3. B 4. B 5. D 6. C 7. A 8. B 9. C 10. D

Ⅲ.1. C 2. A 3. D 4. C 5. B 6. B 7. A 8. D 9. B 10. C

Ⅳ.1. 所有这些都是根据重力分离原理使油滴浮升到板表面,在此进行聚结和油滴捕集。

2. 其主要用途:(1)将平台排放的污水集中到一处;(2)污水可沿此导管排入水下较深位置,不受波浪的影响,即使在异常状况下也不会出现水面油污;(3)万一发生油漫流甲板的事故,可以报警或停机。

3. 排放管桩用于浅水水域时长度和水深一样,在发生意外事故时可使油污容量最大化,使水面污染最小化。

4. 水流流经多组折流板时产生无流动区,缩短了指定油滴与主水流分离所必需上升通过的距离。

5. 根据联邦政府环保条例,游离水脱除器排出的采出水至少要经过某种形式的初步处理后方可排入排放管桩或隔油管桩。

V. 1. 隔油管桩比标准排放管桩能更有效地分离油,而且,从油分离的角度看,隔油管桩还有净化油污砂子的优点。大多数政府都立法规定,采出的砂子必须不含游离油方能排放。使用标准排放管桩时,处理容器排出的油污砂子能否符合这一规定令人怀疑。

2. 甲板污水来源于雨水或洗刷水,通常有油滴分散在富含氧气的淡水或盐水中。水中的氧会使其具有很强的腐蚀性,如果与采出水掺混,会在隔油罐、板式聚结器或浮选池里引起结垢或堵塞。这种污水的排放量很缺乏规律性,因此,容易导致设备出现故障。最终,这些污水在重力作用下会流到一个较低部位集中起来,如果处理设备标高较高,应当用泵;如果处理设备标高较低,可以不用泵。

第3章 天然气形成与集输

3.1 天然气形成与组分(1)

🔲 导语

天然气是亿万年前的有机质降解形成的,以溶解于重烃和水或游离态等形式在一定压力下储存于地壳的岩层中。天然气是烃类气体的混合物,主要成分是甲烷,但通常也含有不同量的其他烷烃等,甚至还有少量的二氧化碳、氮和硫化氢。天然气可大致分为伴生天然气、非伴生天然气和非常规气,以及富气或贫气、湿气或干气等。

🔲 课文

有机质的降解机理主要有两个:生物降解和热降解。生物气是浅层气,温度较低,是厌氧菌分解沉积有机质形成的。相比较而言,热降解在深层发生,其过程可以分为如下两步:(1)有机质裂解为液态和气态的碳氢化合物(和液态烃同时产生的气体称为原生热成因气);(2)高温下液态烃热裂解为气体(次生热成因气)和焦沥青。生物成因气基本都是由甲烷构成,而热成因气还含有大量的乙烷、丙烷、丁烷以及重烃等。

天然气的主要成分是甲烷,其他组分主要是烷烃,如乙烷、丙烷和丁烷等。很多天然气藏还含有氮气、二氧化碳、硫化氢,有的天然气中还可能有微量的氩、氢、氦等。天然气的组分差别巨大。一些情况下,天然气还含有 C_5 以上的碳氢化合物,这些重组分分离后可作为轻质汽油。有时,芳香族化合物如苯、甲苯、二甲苯等也会出现在天然气中。芳香族化合物有毒,给天然气生产输送等带来了安全隐患。一些天然气组分中,汞以蒸气的形式存在于气相中或以有机金属化合物的形式存在于液相中。尽管汞浓度很低,但由于其毒性和腐蚀性强,危害性还是很大的。

当天然气脱除大部分其他常见的伴生烃,组分几乎接近于纯甲烷时,就称为"干气";当其他烃类含量较多时,称为"湿气"。在不同的油田、不同的地层中,天然气的组分是不同的,所以人们设计了标准测试方法来确定天然气的组分以便应用。

天然气来源

形成于地层的天然气构成极其复杂,大致可以分为以下三大类:(1)常规气田的非伴生气;(2)常规油田的伴生气;(3)非常规气藏。非常规气藏包括在低渗透岩层中发现的致密气或称致密砂岩气、煤形成过程中伴生的煤层气、永久冻土带下类似冰结构的天然气水合物以及比常规气藏埋藏更深的深层气。非常规天然气中,对天然气集输工业最为重要的是煤层气。

伴生气

伴生气是与原油伴生的、在原油开采过程中一同采出的气体。原油开采过程中不可能只采油不采气,气体总是会随着井筒压力的下降而从原油中析出。只有极少数油藏不含溶解气。

地层产液到达地面后,在井场附近的油罐中被分离为油流(原油或凝析液)、水流(盐水)和气流,分离后的气流富含天然气凝析液(NGL)。NGL 包括乙烷、丙烷、丁烷、戊烷和相对分子质量更大的重烃(C_6 以上),天然气中的重烃一般称为天然汽油。

天然气油中的 NGL 一般用“gal/ft^3”单位来计量,根据重组分的含量,天然气可以分为富气(大于等于 $5 \sim 6$ gal/ft^3 可采油气量)和贫气(小于 1gal/ft^3 可采油气量)。然而,在天然气处理中,富气和贫气的概念并不是指天然气质量的好坏,而是反映天然气中液态烃相对含量的多少。

伴生气在井筒中可以起到举升原油的效果。所以,人们将气体从环空中注入,通过气举阀将气体从井底附近注进油管。在井口附近,油气水混合物通过接近大气压力的分离器进行分离,这个过程通常分为两步。油水混合物从压力较低的分离器底部进入油罐进行油水分离。分离器中的气体和原油中析出的气体(剩余气)经过处理变成 NGL,产出的 NGL 在处理厂被分离出丙烷和丁烷或者两者的混合物(又称为液化石油气)。丙烷和丁烷分离后,相对分子质量更大的组分变成凝析液,这些凝析液和原油一起产出或者单独产出。此时的气体就是干气了,经过压缩,就可以输入天然气系统,顶替非伴生气。其他气田的天然气经过这个步骤的预处理后也可以进入天然气体系。伴生气经过洁净化处理并控制在一定的压力后,还可以作为现场燃气涡轮机的燃料。涡轮机启动时用主天然气系统中干净的气,如果现场能够用简单的设备从其他处理厂收集并处理低压气体后供涡轮机运行,会更经济一些。

现在各国政府要求企业停止燃烧伴生气,因为人们认为这是在浪费不可再生资源。回注的气体最终还要采出,所以通常有条款规定了什么时候气体能够回注,什么时候用火炬烧掉。

Key to Exercises

Ⅰ. 1. Two main mechanisms (biogenic and thermogenic) are responsible for this degradation. Biogenic gas is formed at shallow depths and low temperatures by the anaerobic bacterial decomposition of sedimentary organic matter. In contrast, thermogenic gas is formed at deeper depths by (1) thermal cracking of sedimentary organic matter into hydrocarbon liquids and gas and (2) thermal cracking of oil at high temperatures into gas and pyrobitumen.

2. The principal constituent of natural gas is methane. Other constituents are paraffinic hydrocarbons such as ethane, propane, and the butanes. Many natural gases contain nitrogen as well as carbon dioxide and hydrogen sulfide. Trace quantities of argon, hydrogen, and helium may also be present.

3. The varieties of gas compositions can be broadly categorized into three distinct groups: (1) nonassociated gas that occurs in conventional gas fields, (2) associated gas that occurs in conventional oil fields, and (3) unconventional gas.

4. Nonassociated gas is produced from geological formations that typically do not contain much, if any, higher boiling hydrocarbons than methane. Nonassociated gas is directly controllable by the producer; one just turns the valves. The gas flows up the well under its own energy, through the wellhead control valves, and along the flow line to the treatment plant.

5. Because it is a waste of a valuable nonrenewable resource. There are often regulatory restrictions on when produced gas can be reinjected, or flared, with an understanding that any reinjected gas must eventually be produced.

Ⅱ. 1. B 2. D 3. C 4. A 5. A 6. B 7. D 8. A 9. A 10. B

Ⅲ. 1. D 2. B 3. A 4. C 5. C 6. D 7. B 8. C 9. D 10. D

Ⅳ. 1. 天然气以溶解于重烃和水或游离态等形式在一定压力下储存于地壳的岩石中。

2. 相比较而言,热降解在深层发生,其过程可以分为如下两步:(1)有机质裂解为液态和气态的碳氢化合物(和液态烃同时生产的气体称为原生热成因气);(2)高温下液态烃热裂解为气体(次生热成因气)和焦沥青。

3. 非常规气藏包括在低渗透岩层中发现的致密气或称致密砂岩气、煤形成过程中伴生的煤层气、永久冻土带下类似冰结构的天然气水合物以及比常规气藏埋藏更深的深层气藏。

4. 然而,在天然气处理中,富气和贫气的概念并不是指天然气质量的好坏,而是反映天然气中液态烃相对含量的多少。

5. 回注的气体最终还要采出,所以通常有条款规定了什么时候气体能够回注,什么时候用火炬烧掉。

Ⅴ.1. 当天然气脱除大部分其他常见的伴生烃,组分几乎接近于纯甲烷时,就称为"干气";当其他烃类含量较多时,称为"湿气"。在不同的油田、不同的地层中,天然气的组分是不同的,所以人们设计了标准测试方法来确定天然气的组分以便应用。

2. 伴生气在井筒中可以起到举升原油的效果。所以,人们将气体从环空中注入,通过气举阀将气体从井底附近注进油管。伴生气经过洁净化处理并控制在一定的压力后,还可以作为现场燃气涡轮机的燃料。涡轮机启动时用主天然气系统中干净的气,如果现场能够用简单的设备从其他处理厂收集并处理低压气体后供涡轮机运行,会更经济一些。

3.2　天然气形成与组分(2)

🔲 导语

伴生气是在常规油气储层中伴随原油生产而产生的天然气,非伴生气是常规气藏中生产的天然气。煤层气是从煤层中开采的,是吸附在煤矿上的天然气。页岩气和致密气因其储层低渗透率特点,流动性较常规储层中的天然气差。和常规油气藏不同,这些天然气通常圈闭在气源岩中。天然气在销售之前,要去酸、添味,保证使用安全。

🔲 课文

非伴生气

地层产出的典型非伴生气(有时称为气井气)中重烃(凝析气)含量小于甲烷含量,非伴生气中有时含有非烃物质,如二氧化碳、硫化氢等。生产过程中,直接调整地面阀门就可以控制非伴生气,可控性较高。气流在地层能量下流出井筒,通过地面控制阀和输气管线进入处理厂。处理时整个流程的温度需要根据管线中的压力降至某一数值,保证重烃组分在该温度和压力下凝结,以便分离。

非常规天然气

非常规天然气指的是那些传统上认为难以开采或成本很高的气源。在此主要涉及三种非常规天然气。

(1) 页岩气是产自页岩地层的天然气,是增长最迅速的天然气形式中的一种。页岩既是天然气的气源岩,也是储层,其特点是渗透率低,页岩气比起常规储层气体在岩石中的流动性更有限。早期的页岩气井是垂直井,目前更多的是水平井,并且需要人工增产措施来生产,如水力压裂。只有具有某种地层特性的页岩才会产生气体。这在很大程度上归因于水平钻井井的重大进展,以及井的增产技术和对这些技术成本效益的完善。而水力压裂是其中最重要的。

(2) 致密气是用于描述低渗透率地层中天然气的一个术语,一般指那些低渗透率的气藏,必须使用增产技术,如水利压裂,才能使天然气流入井筒,产生经济效益。致密气没有明确分类,与常规气、页岩气之间的界限也不甚清晰。

(3) 煤层气是指储存在煤中的甲烷气。通过直井、斜井或者水平井中泵的抽汲,煤层的水压下降,煤层气释放排出。漫长的地质史中,上覆沉积有机质和无机质的堆积,使沉积有机质的埋深不断增加,温度、压力相应提高,有机质慢慢地变成煤,甲烷主要是在成煤过程中形成的。这就是所谓的热成因煤层甲烷气。更多的情况是在浅层未成熟的煤层中,细菌活动(包括大气降水从露头或潜伏露头中带入的细菌)控制了煤层气的生成,这称为晚期生物成因煤层气。

煤形成过程中发生的一系列化学反应均会生成煤层气,这些气体的大部分进入上覆岩层或下伏岩层,最后被煤封闭。常规气藏中气体圈闭在孔隙里,而煤层气则不同。在地层压力下,煤层气实际上是吸附在煤颗粒表面或微孔隙中。由于微孔隙的表面积很大,所以单位体积的煤表面吸附的甲烷要比绝大多数砂岩中的甲烷多。

煤层气的产量与煤化程度以及煤的埋深联系紧密。一般来说,煤化程度越高,埋深越大,生气潜力越大,保存条件越好。因为煤的内表面积很大,每吨煤的孔隙超过10亿立方英尺,和常规的砂岩储层相比,在相同的深度和压力下,煤层的平均储气量是常规砂岩的3倍。要使煤层气释放吸附出来,就必须降低地层压力,因此需要排出煤层中的水。气体从煤表面释放后,通过煤骨架运移到达天然裂缝网络。煤层中的这些天然裂缝通常被称为割理,气体就顺着这些裂缝或割理流入井底。煤层气的成分单一,只含有甲烷,不像常规天然气中含有乙烷等,所以煤层气几乎不需要经过加工处理。煤层气的热值要比常规天然气稍高。

天然气的物理性质和化学性质

天然气无色、无味、无形,密度比空气小。天然气在销售之前,要经过去酸、添味、油气水露点调整等处理,处理后的天然气达到规定的压力、发热量或 Wobbe 指数后方可销售。Wobbe 指数(发热量与密度的比值)表示在一定压力下通过给定的孔径进入燃气设备的热量。

由于管路输送的天然气无色无味,所以通常要求在天然气中加入特定的气味,这样一旦天然气发生泄漏,就可以立即发现。

天然气送达客户之前,通常是在天然气中加入一些有机含硫化合物来达到添味的目的。添加了含硫化合物(一种化学添味剂,即通用化学表达式为 R—SH 的硫醇,有臭鸡蛋气味)的天然气一旦发生泄漏就可以发现。行业标准是当空气中的天然气浓度达到 1% 时,用户通过气味就可以判断出来。天然气的自燃下限是 5%,因此,行业标准数值是此下限值的 1/5,这样微量的添味剂即使燃烧也不会因为硫的存在而造成严重后果或者中毒事件。

Key to Exercises

Ⅰ. 1. Because of the significant advances in the use of horizontal drilling and well stimulation technologies and refinement in the cost – effectiveness of these technologies.

2. Coal bed methane is the generic term given to methane gas held in coal and released or produced when the water pressure within the buried coal is reduced by pumping from either vertical or inclined to horizontal surface holes.

3. The methane is predominantly formed during the coalification process whereby organic matter is slowly transformed into coal by increasing temperature and pressure as the organic matter is buried deeper and deeper by additional deposits of organic and inorganic matter over long periods of geological time. This is referred to as thermogenic coal bed methane. Alternatively, and more often (but not limited to) in lower rank and thermally immature coals, recent bacterial processes (involving naturally occurring bacteria associated with meteoric water recharge at outcrop or subcrop) can dominate the generation of CBM. This is referred to as late – stage biogenic coal bed methane.

4. Natural gas is colorless, odorless, tasteless, shapeless, and lighter than air.

5. The natural gas after appropriate treatment for acid gas reduction, odorization, and hydrocarbon and moisture dew point adjustment would then be sold within prescribed limits of pressure, calorific value, and possibly *Wobbe index*.

Ⅱ. 1. B 2. A 3. D 4. B 5. A 6. B 7. A 8. C 9. B 10. A

Ⅲ. 1. A 2. D 3. C 4. B 5. C 6. D 7. C 8. D 9. B 10. B

Ⅳ. 1. 早期的页岩气井是垂直井,目前更多的是水平井,并且需要人工增产措施来生产,如水力压裂。

2. 这在很大程度上归因于水平钻井的重大进展,以及井的增产技术和对这些技术成本效益的完善。

3. 煤层气的产量与煤化程度以及煤的埋深联系紧密。一般来说,煤化程度越高,埋深越大,生气潜力越大,保存条件越好。

4. 因为煤的内表面积很大,每吨煤的孔隙超过 $10 \times 10^8 \, ft^3$,和常规的砂岩储层相比,在相同的深度和压力下,煤层的平均储气量是常规砂岩的 3 倍。

5. 天然气在销售之前,要经过去酸、添味、油气水露点调整等处理,处理后的天然气达到规定的压力、发热量或 Wobbe 指数后方可销售。

Ⅴ. 1. 致密气是用于描述低渗透率地层中天然气的一个术语,一般指那些低渗透率的气藏,必须使用增产技术,如水利压裂,才能使天然气流入井筒,产生经济效益。致密气没有明确分类,与常规气、页岩气之间的界限也不甚清晰。

2. 由于管路输送的天然气无色无味,所以通常要求在天然气中加入特定的气味,这样一旦天然气发生泄漏,就可以立即被发现。天然气送达客户之前,通常是在天然气中加入一些有机含硫化合物来达到添味的目的。添加了含硫化合物的天然气一旦发生泄漏就可以发现。

3.3　天然气集输(1)

导语

从气井采出的天然气称为井口气,必须运到加工厂处理后再输送给用户。因为密度低,天然气很难储存或用车辆运输。天然气管道运输既经济,效率又高,然而用于海洋运输不实际。海洋运输天然气通常用液化气船,短距离运输液化或压缩天然气常用油罐车。目前正在开发使用压缩气船进行天然气海上运输,这在某些情况下可以比液化气运输更好。

课文

由于储存困难,天然气在采出之后必须立即输送至一定的区域,从油气田到销售市场可供选择的输送方法很多,包括管道输送、液化天然气(LNG)输送和压缩天然气(CNG)输送等。

管道

天然气在从气源到终端用管道输送非常便利,但不够灵活。由于气体不容易储存,除非适当地增加管线压力,否则管线出现故障,气井、集气站和处理厂都得停工。

过去十年中,全世界平均每年新建 12000mi 的输气管线,其中绝大多数是跨国的。从长远来看,只要政局稳定,管线输送是最好的途径,比如建议修建从阿曼到印度的深水管线,但修建这样一条管线所需的费用尚不清楚。目前认为修建超过 2000mi 的海底管线不经济,因为海底管线的敷设和维护费用相当高昂,而且沿管线进行再压缩非常困难,至今还没有很好的解决方法。如果技术和经济问题得到解决,这种海底管线输送将非常高效。

液化天然气

LNG 是液态的天然气。天然气在大约 −162℃ 时开始液化,液化后的体积只有室温下体积的 1/600。但是,天然气的液化需要复杂的可移动设备,还需要有制冷功能的货轮来运输。自 20 世纪 80 年代中期以来,由于热动力学的发展,液化天然气工厂的建造成本下降,这使得 LNG 技术成为备受世界各国青睐的天然气输送方式。现在,许多 LNG 工厂扩张规模,更多的工厂正在建设。

储存液化天然气需要大型低温储罐,一般储罐直径为 70m,高 45m,能储存 100000m³ 的液化天然气。在客户终端,也需要建立对 LNG 气体进行再处理的配套基础设施,这些设施昂贵且难以维护。

目前,世界上最大的低温冷藏罐能容纳 135000m³ 液化天然气,相当于 $28.6 \times 10^8 ft^3$ 的天然气,但需要的费用很高。LNG 技术是基于大批量输送和连续运行来提高效率从而降低成本的,对于孤立的储层(如海上)或者需求较小的用户,LNG 技术的推广应用难度较大。因此,间歇供气的小型气源对于 LNG 销售商没有吸引力。现在正在考察一种密封隔热性能良好的 LNG 储罐商业运行的可行性,如果可行,小批量的液化天然气也可以从储存地向外运输,就像现在的油罐车模式。即便如此,但要保证液化天然气在没有蒸发逸失的情况下储存一定时间(几个月),难度是很大的。

压缩天然气

高度压缩的天然气可以储存在油罐中。一般来讲,富气(富含乙烷、丙烷等)压力为 1800 psi,贫气(主要为甲烷)大概为 3600 psi,这些压力下的气体可以称为压缩天然气。一些国家将压缩天然气(CNG)用于交通,以替代常规燃料(如汽油或柴油)。加气站可以通过管线来供应,但能将天然气加压到 3000 psi 的压缩机的售价、维护和运行费用都是非常高昂的。

另外一种方法是用货轮直接装载大直径的长形储罐,但气体必须经过干燥、压缩和冷却。在精确的温度控制下,根据货轮的负载能力(即容积限制、管线材料的压力安全限制),尽可能多地输送气体。此外,还需要压缩机与冷却机,但这些普通的设备要比液化机便宜得多,能大大降低成本。支持者还认为,终端装置较简单,因此成本要低一些。两家公司对各自的新型CNG输送方式作了改进,如下所述。

Votrans 是 EnerSea 运输公司开发的一种新型 CNG 海洋运输技术。工程研究表明,使用这项技术每艘货轮可以将 $2\times10^8\text{ft}^3$ 的 CNG 输送 4000mi,而且成本远低于 LNG。这项技术需要大量大直径管汇叠置,实际上相当于海上可移动的管道。

为了维持温度,这些管线都封存在充满氮气、绝热的容器内。这项技术储存效率要高于 LNG,所需压力更低(只有 LNG 的 40%),储存量更大,成本也较低,而且对贫气和富气都适用。在输送加工过程中,Votrans 技术将气体逸失量从 LNG 技术的 20% 降低到 7%。

Coselle 是 Cran & Stenning 技术公司开发的 CNG 运输技术。该系统采用常规的长为 10.6mi、直径为 6in、壁厚 1/4in 的大型盘管。整个流程共有 108 根盘管,输送能力为 $330\times10^6\text{ft}^3$,压力 3000 psi 下储气温度为 50 ℉。美国航运局/船级社和挪威船级社对这项技术的评定为"至少和常规输气一样安全"。船只可以在较简便的海上设施上装载,比如海上锚系装置,这种装置通常与岸上或者平台上的压缩机站的软管相连接。Coselle 和 Votrans 也许会即将成为 CNG 高压集输的商业化方法。

CNG 技术使天然气短程输送成为可能。这项技术旨在使一些由于输送技术的局限或者 LNG 过于昂贵而没能开发的海上油田具备商业价值。从技术的角度讲,CNG 对设备和基础设施的要求较低。

结果表明,在 2500mi 范围内 LNG 输送天然气的成本为 15~25 美元/10^6Btu,而 CNG 输送成本为 0.93~2.23 美元/10^6Btu;在 2500mi 范围外,由于 CNG 输送量要比 LNG 小,其输送成本要略高于 LNG。

Key to Exercises

Ⅰ.1. The most cost-effective way for natural gas transportation is by pipelines.

2. LNG will likely play an increasing role in the development of giant gas fields, as most countries, especially net oil importers, are keen on developing their gas reserves, however stranded, for greater energy independence and ex-

tending domestic oil reserves where applicable, as well as for environmental reasons.

3. Gas can be transported in containers at high pressures, typically 1800 psig for a rich gas (significant amounts of ethane, propane, etc.) to roughly 3600 psig for a lean gas (mainly methane). Gas at these pressures is termed *compressed natural gas.*

4. Compressed natural gas is used in some countries for vehicular transport as an alternative to conventional fuels (gasoline or diesel). The filling stations can be supplied by pipeline gas, but the compressors needed to get the gas to 3000 psig can be expensive to purchase, maintain, and operate. An alternative approach has dedicated transport ships carrying straight long, large – diameter pipes in an insulated cold storage cargo package.

5. " Coselle " CNG technology is from Cran & Stenning Technology Inc. The system uses conventional, 10. 6 – mile – long, 6 – in diameter, 1/4 – in wall thickness line pipe in large coils (coselles). Such a CNG carrier may have 108 coselles with a 330 – MMcfg capacity. Stored gas temperature is 50 ℉ at 3000 psi. These ships can be loaded at relatively simple marine facilities, including offshore buoy moorings, through flexible hoses connected to onshore or on – platform compressor stations.

Ⅱ. 1. C 2. B 3. A 4. D 5. C 6. A 7. B 8. A 9. D 10. B

Ⅲ. 1. D 2. D 3. C 4. B 5. C 6. A 7. C 8. B 9. D 10. C

Ⅳ. 1. 船只可以在较简便的海上设施上装载,比如海上锚系装置,这种装置通常与岸上或者平台上的压缩机站的软管相连接。

2. 由于气体不容易储存,除非适当地增加管线压力,否则管线出现故障,气井、集气站、处理厂都得停工。

3. 目前认为,修建超过 2000mi 的海底管线是不经济的,因为海底管线的敷设和维护费用相当高昂,而且沿管线进行再压缩非常困难,至今还没有很好的解决方法。

4. 自 20 世纪 80 年代中期以来,由于热动力学的发展,液化天然气工厂的建造成本下降,这使得 LNG 技术成为备受世界各国青睐的天然气输送方式。现在,许多 LNG 工厂扩张规模,更多的工厂正在建设。

5. CNG 技术旨在使一些由于输送技术的局限或者 LNG 过于昂贵而没能开发的海上油田具备商业价值。

Ⅴ. 1. LNG 技术是基于大批量输送和连续运行来提高效率从而降低成

本的,对于孤立的储层(如海上)或者需求较小的用户,LNG 技术的推广应用难度较大。因此,间歇供气的小型气源对于 LNG 销售商没有吸引力。现在正在考察一种密封隔热性能良好的 LNG 储罐商业运行的可行性,如果可行,小批量的液化天然气也可以从储存地向外运输,就像现在的油罐车模式。

2. 另外一种方法是用货轮直接装载大直径的长形储罐,但气体必须经过干燥、压缩和冷却。在精确的温度控制下,根据货轮的负载能力(即容积限制、管线材料的压力安全限制),尽可能多地输送气体。此外,还需要压缩机与冷却机,但这些普通的设备要比液化机便宜得多,能大大降低成本。

3.4 天然气集输(2)

导语

除了管道输送、液化天然气(LNG)输送和压缩天然气(CNG)输送方式外,天然气还可以通过其他许多形式输送,比如,固化输送(GTS)——将天然气转化为水合物;动力化输送(GTP)——将天然气转化为电力;气体液化输送(GTL)——将天然气转化为大量的其他产品,如洁净的燃料、塑料原料或甲醇等;以及商品化输送(GTC)——将天然气转化为各种商品,如铝、玻璃、水泥、铁等。

固化输送(GTS)

气体可以在固化后进行输送,即所谓的天然气水合物。天然气水合物是天然气与水形成的冰状物质。关于天然气水合物输送的研究仍然处于试验阶段,但普遍认为这项技术将取代液化天然气或管道输送。

天然气的固化包括三个步骤:生产、运输、再汽化。天然气水合物是天然气中的小分子如甲烷、乙烷、丙烷在水的氢键作用下形成的稳定的三维笼状结构物质。若干水分子在氢键的作用下形成笼子,气体分子则被圈在笼中。当压力高于气水两相相图中的平衡线而温度低于此平衡线时,天然气与水形成雪花状的水合物。

在石油天然气工业中,水合物对管道输送危害极大,在输送过程中必须非常小心,防止水合物生成。如果未采取预防措施,如注入甲醇,管线就可能堵塞。然而,在永久冻土带和海底以下 500m 处发现了大量的天然气水合物,如果开发方式得当,天然气水合物将会成为未来 30 年的主要能源。

为了便于运输,可在 80 ~ 100bar 的压力、2 ~ 10℃时使天然气和水形成水合物。如果在 −15℃左右冷藏,即便在大气压下水合物分解也十分缓慢,这样就可以用隔热性能非常好、接近绝热的状态下的简便储气罐运输。在目的地,通过控制升温,水合物融化,分解成水和气,气体经过干燥处理后就可用于发电或其他用途。每吨水合物能释放出大约160m³天然气,这取决于加工处理过程。天然气在经过简单的处理(如净化)后方能进行水合物生产。陆地上是通过一些可移动的设备,海上可使用浮动的生产、储存和卸载船来实现,这种方法具有很大的商业价值。

水合物的生产看起来似乎就是将凉水和天然气混在一起而已,实际上,整个过程是在加工中经过一系列转换才完成的,每一次都进一步压缩水合物。制成的水合物被储存最后装载到货轮,到了目的地后,水合物分解利用。如果目的地需要水,从水合物中分离出来的水可以派上用场,否则就作为压舱物返回;分离出来的水饱和了气体,对气体的再溶解能力较小。

在1~10atm下,水合物可以在常温下储存(-10~0℃),此时1m³水合物可以容纳大概160m³气体,这还是比较有吸引力的。虽然压缩天然气在3000 psi的高压下可以容纳200m³的气,液化天然气在-162℃的低温下可以容纳637m³气体,但天然气水合物容易制成,而且更安全、更经济。

水合物储气在低温下更容易实现,即低温下单位体积水合物所含有的气体比压力下降时自由状态或压缩状态下所含有的气体要多。在输送环境较恶劣的情况下,和管道输送与液化天然气输送相比,水合物输送的成本较低。由于水合物储存与输送不需要低温或高压,因此较有前景。每立方米液化天然气含637m³天然气,而水合物只有160m³。单独来看这是水合物的劣势,但考虑到水合物运输成本较低,所以水合物还是比较经济的。

动力化输送(GTP)

目前输送的天然气多用来发电。可以在距离气源较近的地方发电并用电缆输送到用户,这就是气转电。例如,可在相对平静的水域建设海上发电厂,利用海上天然气发电,供给陆地或者海上用户使用。但是,在海上架设高压电缆的成本几乎和敷设海底管线的成本一样高,而气转电的初衷是降低运输成本,因此这种做法不太可行。

还有一些实际的问题需要考虑。比如对于伴生气,如果除了转化为电力以外没有其他处理途径,那么一旦发电机不能工作,整个采油系统都必须停工,或者把伴生气排放到大气中燃烧。如果发电厂的设备出现了故障,必须尽快关闭(在60s之内),防止问题进一步扩大。然而,为保证关闭系统的安全性,结构复杂的系统在关闭前还需要冷却循环或者要进行洁净化处理,显然这不适合快速关闭。最后,如果发电厂不能说关就关,而且在很短时间(1h)内就能启动,那么操作人员如果考虑到电网售电方的经济索赔,就不敢轻易停产。

气体液化输送(GTL)

GTL 运输是将天然气转化为液体,比如合成原油、甲醇和氨,然后再运输。GTL 并不是新技术。首先在甲烷中混入蒸气,在适当的催化剂作用之下可以生成合成气(一氧化碳和氢气的混合物);然后把合成气转化为液体;产出液可以是燃料,通常是洁净的动力燃料(合成石油)或润滑剂、氨、甲醇、制造塑料的原料。人们对这个复杂、能量密集型的过程进行了多次改进,已经产生了成百上千的专利。近年来的改进主要集中在降低成本、减少过程能耗,尤其是催化剂的合理使用与氧气的加入方式上。

Key to Exercises

Ⅰ. 1. It is the product of mixing natural gas with liquid water to form a stable water crystalline ice – like substance. The solid has a snow – like appearance.

2. Natural gas hydrates are created when certain small molecules, particularly methane, ethane, and propane, stabilize the hydrogen bonds within water to form a three – dimensional, cage – like structure with the gas molecule trapped within the cages. A cage is made up of several water molecules held together by hydrogen bonds. Hydrates are formed from natural gas in the presence of liquid water, provided the pressure is above and the temperature is below the equilibrium line of the phase diagram of the gas and liquid water.

3. For gas transport, natural gas hydrates can be formed deliberately by mixing natural gas and water at 80 to 100 bar and 2 to 10℃. If the slurry is refrigerated to around − 15℃, it decomposes very slowly at atmospheric pressure so that the hydrate can be transported by ship to market in simple containers insulated to near – adiabatic conditions.

4. Because installing high – power lines to reach the shoreline appears to be almost as expensive as pipelines, that gas to power could be viewed as defeating the purpose of an alternative less expensive solution for transporting gas.

5. It can be a fuel, usually a clean – burning motor fuel (syncrude) or lubricant, or ammonia or methanol or some precursor for plastics manufacture. Other GTL processes are being developed to produce clean fuels, e. g. , syncrude, diesel, or many other products, including lubricants and waxes, from gas but require a complex (expensive) chemical plant with novel catalyst

technology.

Ⅱ. 1. C 2. B 3. B 4. C 5. A 6. D 7. A 8. C 9. D 10. B

Ⅲ. 1. C 2. D 3. A 4. B 5. D 6. C 7. A 8. C 9. D 10. A

Ⅳ. 1. 关于天然气水合物输送的研究仍然处于试验阶段,但普遍认为这项技术将取代液化天然气或管道输送。

2. 在石油天然气工业中,水合物对管道输送危害极大。

3. 在输送过程中必须非常小心,防止水合物生成。如果未采取预防措施便注入甲醇,管线就可能堵塞。

4. 然而,在永久冻土带和海底以下 500m 处发现了大量的天然气水合物,如果开发方式得当,天然气水合物将会成为未来 30 年的主要能源。

5. GTL 的另一用途是生产清洁燃料,比如合成原油、柴油或者其他产品,包括润滑油、蜡。但是生产这些产品的化工厂需要复杂、昂贵的设备以及过硬的催化技术。

Ⅴ. 1. 由于在海上架设高压电缆和敷设海底管线的成本一样高,而气转电的初衷是降低运输成本,因此这种做法不太可行。在长距离电缆输送电力的过程中,能量损耗非常大。流电的输送比直流电更耗能。此外,在交流电转变为直流电时,要消耗能量;将输送时的高压转换为用户需求的低压电时同样有损耗。

2. 如果只用伴生气发电,那么一旦发电机不能工作,整个采油系统都必须停工,或者把伴生气排放到大气中燃烧。如果发电厂的设备出现了故障,必须尽快关闭,防止问题进一步扩大。然而,为保证关闭系统的安全性,结构复杂的系统在关闭前还需要冷却循环或者要进行洁净化处理,显然这不适合快速关闭。最后,如果发电厂不能说关就关,而且在很短时间内就能启动,那么操作人员如果考虑到电网售电方的经济索赔,就不敢轻易停产。

第4章 天然气分离

4.1 天然气处理的基本概念

📭 导语

　　天然气处理虽然在许多方面比原油处理要简单,但是天然气也必须经过处理才能最终使用。消费者使用的天然气几乎都由甲烷构成。无论天然气源于何处(油藏、气藏或凝析气藏),一般都混有其他烃类,主要是乙烷、丙烷、丁烷和戊烷,此外还含有水蒸气、硫化氢、二氧化碳、氦、氮及其他组分。天然气处理就是要生产出符合管道运输标准的干气。

📭 课文

　　经由油气田集气管网来的原料天然气必须经过处理后,才能进入长输管道供消费者使用。天然气处理的目的是分离出天然气、凝析油、非凝析物和酸性气体以及来自生产天然气井的水,达到成品销售或废物处置的标准。图4.1.1给出了典型的处理流程模块。每个模块包括一个或一组具有特定作用的设备。图中的所有模块并不一定每个处理厂都会用到。在某些情况下,只需要对天然气进行简单的处理;然而,多数情况下,天然气要通过处理

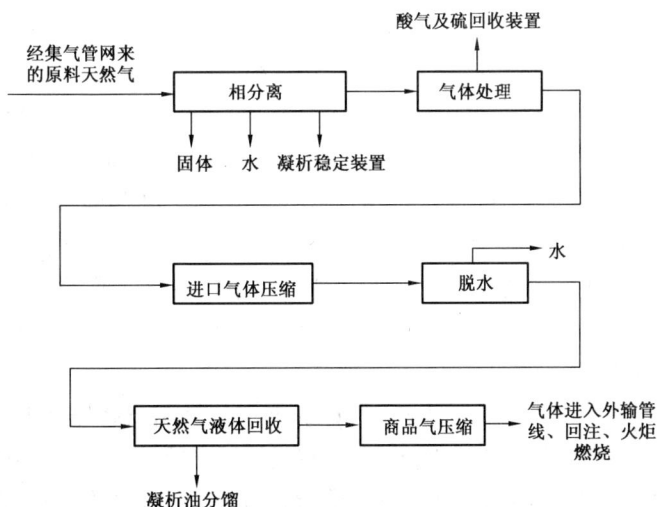

图 4.1.1　典型的简化陆上处理工艺过程

厂处理,去除杂质、水和多余的烃液并进行外输压力控制。在具体应用中,流程可能和图4.1.1中的顺序不同,虽然这个流程是常用的。在气田开发方案设计阶段就确定选用什么样的模块和流程顺序。

处理模块

第一个模块是分相物理分离模块,主要是将不同相的气体、液态烃、液态水或固体分离。湿气的相分离通常在进口分离器内完成。进气端的情况很复杂,给处理厂送气的集气管道中流体一般都为两相或三相,因此常有液体段塞流。段塞通常是由于集气管道高度变化、气体流量变化、输送过程中温度和压力变化引起的。在稳态条件下,如果流动状态选择不当,段塞流也可能在水平管道中出现。生产和处理设备中出现段塞,会对生产设施的运作产生负面影响,引起机械问题(由于高速和高动量)和处理过程问题(提高液面,引起浪涌和滑脱)。

气体管道中使用段塞捕集器来分散液体段塞的能量,减少紊流,降低气体和液体的流速,使流态保持层流,产生重力分离。段塞捕集器用来分离凝析烃、入口水和气体。段塞捕集器分离出的液体进入三相分离器中,分离出两相液体、凝析烃和水/甲醇或水/乙二醇并产出。从三相分离器分离出的塔顶气需要作燃料气时,需要再次压缩。

从天然气中回收的凝析烃不经过处理也可以运输,但这样的凝析液含有大量的甲烷和乙烷,很容易在储存罐中挥发,所以一般需要经过稳定处理,把凝析液中的轻组分完全去除,以符合安全运输。稳定处理过程通常通过蒸馏完成,从凝析液中完全去除轻烃。稳定处理后的液体要符合蒸汽压指标(雷德蒸汽压小于10psi),因为要注入有一定压力限制的管道或运输压力罐。

天然气处理的下一步是酸性气体的处理。除了重烃和水蒸气,天然气还包含其他必须除去的污染物。二氧化碳、硫化氢和其他含硫成分如硫醇,都是需要全部或部分去除的化合物。这些化合物统称为酸性气体。硫化氢和水混合时形成弱硫酸,而二氧化碳和水混合后形成碳酸,因此称为酸性气体。含硫化氢和其他含硫化合物的天然气叫做"酸气",而只含有二氧化碳的气体叫做"甜气"。含二氧化碳或硫化氢都是不符合要求的,因为这些物质能引起腐蚀,造成严重的安全隐患。

根据处理厂进口压力的大小,处理的下一步是将气体增压到门限压力,一般是$300 \sim 400psi$,或是控制露点,回收天然气凝析液。露点控制要符合规范指标,避免水合物形成。在天然气管道运输中,工程师主要担忧的是天然气水合物的形成,因为水合物能引起管道堵塞和其他相关问题。在处理厂中,阻止水合物形成的方法包括使用化学抑制剂来降低水合物的形成温度,或脱水除去可能形成水合物的水。

烃露点控制或液态烃回收涉及气体冷却和液体凝结析出两个过程。烃露点的控制既可以通过脱水然后冷却/凝析完成,也可以通过抑制/冷却/凝析共同作用完成。气体经过阀门降压后可以自动冷却,通过外部制冷设备也可以完成冷却。气体冷却的温度取决于处理目的,销售用气态烃露点标准和大量回收液烃的温度标准是不同的。下面三种情况下,希望实现液烃的最大收率。第一种情况是当处理伴生气时,期望拥有最大的凝析液产量;第二种情况是处理反凝析气时,目的是获取凝析液并将气体重新回注地层;第三种情况是在某些市场中,从凝析油中产出的天然气凝析液(NGL)作为液体比作为销售用气体更有经济价值。是把凝析液留在气流中(但也要达到露点要求)还是将其当作液体回收,纯粹是由经济因素决定的,即对比作为燃料还是液体化学原料哪个用途的价值更大。如果液体价值低于气体,凝析液应尽可能地留在气体中;反之,应最大程度回收凝析液。

如果出厂前气体压力低于销售管道压力(大约 700~1000psi),就要增压到管道压力。为了减少管道的直径,气体在高压下运输。管道在高压下(1000psi 以上)运行还可以保持气体处于致密压缩状态,从而防止发生凝析和两相流动。通常需要两到三级压缩达到销售气体压力要求。

如果没有气体管道,分离出的伴生气可能经火炬燃烧。火炬燃烧法是否可行取决于当地法规及油气田所处位置。这种情形更多的做法是将分离气经压缩后回注产气层保存,以后再开采销售。在凝析气藏中,气体常用来回注,以提高珍贵的液态烃采收率。

Key to Exercises

I. 1. The objective of gas processing is to separate natural gas, condensate, noncondensable, acid gases, and water from a gas – producing well and condition these fluids for sale or disposal.

2. The first unit module is the physical separation of the distinct phases, which are typically gas, liquid hydrocarbons, liquid water, and/or solids.

3. In addition to heavy hydrocarbons and water vapor, natural gas often contains other contaminants that may have to be removed. Carbon dioxide (CO_2), hydrogen sulfide (H_2S), and other sulfur – containing species such as mercaptans are compounds that require complete or partial removal. These compounds are collectively known as "acid gases."

4. Three situations motivate maximum condensate recovery. The first is the desire to maximize condensate production when processing associated gas. The second situation occurs when processing retrograde condensate gas; here the objective is to recover the condensate and reinject the gas into the formation. Third, in some markets the natural gas liquids (NGLs) produced from the condensate may be more valuable as liquid products than as sales gas components, i. e. , their recovery will yield a better profit.

5. If gas is produced at lower pressures than typical sales pipeline pressure (approximately 700 - 1000 psig), it is compressed to sales gas pressure. Compression typically requires two to three stages to attain sales gas pressure.

Ⅱ. 1. C 2. A 3. B 4. D 5. A 6. B 7. C 8. A 9. D 10. C

Ⅲ. 1. C 2. A 3. C 4. B 5. B 6. C 7. B 8. B 9. A 10. D

Ⅳ. 1. 第一个模块是分相物理分离模块,主要是将不同相的气体、液态烃、液态水或固体分离。

2. 生产和处理设备中出现段塞,会对生产设施的运作产生负面影响,引起机械问题(由于高速和高动量)和处理过程问题(提高液面,引起浪涌和滑脱)。

3. 但这样的凝析液含有大量的甲烷和乙烷,很容易在储存罐中挥发。

4. 天然气管道运输中,工程师主要担忧的是天然气水合物的形成,因为水合物能引起管道堵塞和其他相关问题。

5. 管道在高压下运行还可以保持气体处于致密压缩状态,从而防止发生凝析和两相流动。

Ⅴ. 1. 经由油气田集气管网来的原料天然气必须经过处理后,才能进入长输管道供消费者使用。天然气处理的目的是分离出天然气、凝析油、非凝析物和酸性气体以及来自生产天然气井的水,达到成品销售或废物处置的标准。在某些情况下,只需要对天然气进行简单的处理;然而,多数情况下,天然气要通过处理厂处理,去除杂质、水和多余的烃液并进行外输压力控制。

2. 烃露点控制或液态烃回收涉及气体冷却和液体凝结析出两个过程。烃露点的控制既可以通过脱水然后冷却/凝析完成,也可以通过抑制/冷却/凝析共同作用完成。气体经过阀门降压后可以自动冷却,通过外部制冷设备也可以完成冷却。气体冷却的温度取决于处理目的,销售用气态烃露点标准和大量回收液烃的温度标准是不同的。是把凝析液留在气流中(但也要达到露点要求)还是将其当作液体回收,纯粹是由经济因素决定的,即对比作为燃料还是液体化学原料哪个用途的价值更大。如果液体价值低于气体,凝析液应尽可能地留在气体中;反之,应最大程度回收凝析液。

4.2 重力分离

导语

物理分离液体、气体或固体常用动量学、重力沉降和聚结三个原理,其中重力是完成分离的主力。重力分离器是可将混合相分离为各自相对独立的气相和液相的压力容器。重力分离器通常从几何结构上分为垂直分离器和水平分离器,从作用上分为两相分离器和三相分离器。当气液流速比非常高时,分离器有时也称为净气器。

课文

完成物理分离液体、气体或固体的三个原理是动量学、重力沉降和聚结。任何一台分离器都会使用这三个原理中的一个或多个。要进行分离,液相必须是不相溶的并且具有不同的密度。动量产生的力改变液体的运动方向,常用于液相的大体积粗分离。重力可以减小液体的速度,使液滴沉降分离在给定的空间内。重力是完成分离最主要的力,最重的液体沉降到底部,而最轻的升到顶部。然而,非常小的液滴(如雾流)很难靠重力分离。这些液滴可以聚结成大液滴,然后再靠重力分离。

重力分离器

重力分离器是可将混合相分离为各自相对独立的气相和液相的压力容器。在重力分离器中,重力控制分离,降低气体的速度可以提高气液分离的效率。重力分离器通常根据几何结构(垂直、水平)和作用分类(两相分离器、三相分离器)。当气液流速比非常高时,分离器有时也称为净气器。这些容器通常有一个很小的液体收集区,并且只在以下情况推荐使用:

(1)进行二次分离,去除来自处理设备(吸收器,液体除尘器)的携带液。

(2)当管道不是很长时,对输气管道分离器下游的流体进行分离。

(3)气液比非常高时,进行混杂分离。

净气器的配置和尺寸与普通分离器的要求相同。

重力分离器组成与特点

所有重力分离器通常都有如下结构和特点：

(1)带入口分流器的主要气液分离段，用于除去气体中大部分液体。

(2)提供足够滞留时间的重力沉降分离段，以便恰当地沉降。

(3)在气体出口设置除雾器，用于捕捉夹带液滴和太小而无法重力沉降的液滴。

(4)适当的压力和液面控制。

重力分离器是设计成水平的或垂直的压力容器。图4.2.1是典型的三相水平分离器的示意图。

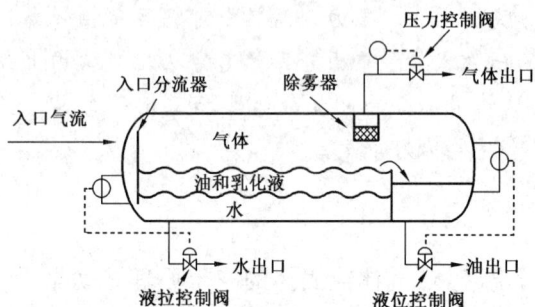

图 4.2.1　水平三相分离器的典型示意图

液体进入分离器并撞击到进口分流器。突然的动量变化导致大部分液体和气体初步分离。在多数分设计中，入口分流器带有一个降液管，引导液体在油水界面下流动。这迫使入口的油水混合物与容器底部的连续水相混合，并且上升到油水界面上部，这个过程叫"水洗"，能促进夹带在连续油相中的水滴聚集。入口分流器确保液体中基本不含气体，并且水洗确保液体不落在油气或油水界面上部，否则与容器内液体混合后使油水界面很难控制。

容器内集液段提供足够时间，使油和乳化液在顶部形成"油垫"。自由水沉降在底部，产出水从容器的挡油板上游的喷嘴流出。界面调节器能感应油水界面的高度，调节器向排水阀发出信号，排出定量的水，使油水界面保持在设计高度上。气流水平流动并通过除雾器到达控压阀，控压阀保持管内恒定压力。

图4.2.2显示了垂直三相分离器的典型构造。如同水平分离器一样，在垂直分离器中，流体从侧面进入容器，并且入口分流器将大部分气体分离。气体向上移动，穿过除雾器除去悬浮的雾流，然后干气流出。降液管输送从气油界面收集到的液体，以避免干扰撇油。

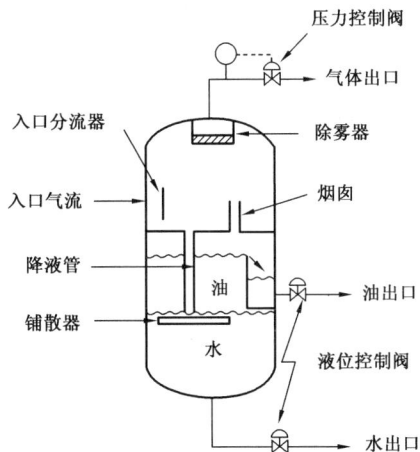

图 4.2.2　垂直三相分离器的典型示意图

烟囱用来平衡下部和气体区的压力。降液管的出口或铺散器位于油水界面处。当油在这里上升时,任何油相中的自由水都将分离出来。水滴流向与油流相反。同理,当水向下流时,束缚在水相中的油滴逆流向上。

三相垂直分离器的操作原理与前面描述的三相水平分离器相同,本质上唯一不同的是,水平分离器是切线流,而垂直分离器是平行流。在垂直分离器中,液位控制也不是很关键的,在不影响操作效率的情况下,液位可以在几英寸的范围内波动,然而液位却能影响降液管的压降,因此影响除雾设备的排驱。

分离器的选择

分离器的选择没有简单的规则。有时,需要对这两种构造进行评价来决定哪一种更加经济可行。

水平分离器

水平分离器常用于如下情况:

(1)大容量的气体和液体。

(2)较高或中等气油比气流。

(3)泡沫油。

(4)三相分离。

水平分离器的优点:

(1)和垂直容器相比,处理同样的天然气量,用小直径的就可以。

(2)没有逆流(气体流向与除雾器的排驱方向不相反)。

(3)液体表面大,泡沫扩散充分,一般能降低紊流。

(4)具有较大的缓冲容积。

水平分离器的缺点：

(1)仅仅一部分外壳可用于气体通道。

(2)占地面积大(除非叠放)。

(3)液位控制很关键。

(4)更难进行产出砂、泥浆、蜡和石蜡等的清理。

垂直分离器

垂直分离器用于如下情况：

(1)小流量的气体或是液体。

(2)高气油比或总气量很低。

(3)空间受限。

(4)容易控制液面。

垂直分离器的优点：

(1)液位控制不是很关键。

(2)拥有很好的底部排泄口和清洁设备。

(3)能很方便处理更多的砂、泥和石蜡等。

(4)重新被夹带走的倾向很小。

(5)足够大的尺寸使气体流到上部,油流到下部。

(6)占用空间小。

垂直分离器的缺点：

(1)处理一定的气量需要更大尺寸的直径,在非常低的气油比或非常高的气油比以及设备清洗方面有更强的竞争性。

(2)有可能产生液流段塞时不推荐使用。

(3)顶部安装的器具和安全设备难于维修保养。

Key to Exercises

Ⅰ.1. Three principles used to achieve physical separation of gas and liquids or solids are momentum, gravity settling, and coalescing.

2. They are recommended only for the following items.

(1) Secondary separation to remove carryover fluids from process equipment such as absorbers and liquid dust scrubbers.

(2) Gas line separation downstream from a separator and where flow lines are not long.

(3) Miscellaneous separation where the gas – liquid ratio is extremely

high.

3. All gravity separators normally have the following components or features.

（1）A primary gas/liquid separation section with an inlet divertor to remove the bulk of the liquid from the gas.

（2）A gravity – settling section providing adequate retention time so that proper settling may take place.

（3）A mist extractor at the gas outlet to capture entrained droplets or those too small to settle by gravity.

（4）Proper pressure and liquid – level controls.

4. Essentially, the only difference is that horizontal separators have separation acting tangentially to flow, whereas vertical separators have separation acting parallel to flow. In the vertical separator, level control is not also critical, where the liquid level can fluctuate several inches without affecting operating efficiency. However, it can affect the pressure drop for the downcomer pipe （from the demister）, therefore affecting demisting device drainage.

5. Horizontal separators are used most commonly in the following conditions.

（1）Large volumes of gas and/or liquids.

（2）High – to – medium gas/oil ratio （GOR） streams.

（3）Foaming crudes.

（4）Three – phase separation.

Ⅱ. 1. A 2. A 3. C 4. B 5. C 6. B 7. C 8. D 9. D 10. A

Ⅲ. 1. D 2. C 3. B 4. A 5. D 6. C 7. A 8. C 9. B 10. D

Ⅳ. 1. 在重力分离器中,重力控制分离,降低气体的速度可以提高气液分离的效率。

2. 调节器向排水阀发出信号,排出定量的水,使油水界面保持在设计高度上。

3. 如同水平分离器一样,在垂直分离器中,流体从侧面进入容器,并且入口分流器将大部分气体分离。

4. 三相垂直分离器的操作原理与前面描述的三相水平分离器相同。

5. 在垂直分离器中,液位控制也不是很关键的,在不影响操作效率的情况下,液位可以在几英寸范围内波动。

Ⅴ. 1. 完成物理分离液体、气体或固体的三个原理是动量学、重力沉降、聚结。任何一台分离器都会使用这三个原理中的一个或多个。要进行分离，液相必须是不相溶的并且具有不同的密度。动量产生的力改变液体的运动方向，常用于液相的大体积粗分离。重力可以减小液体的速度，使液滴沉降分离在给定的空间内。重力是完成分离最主要的力，最重的液体沉降到底部，而最轻的升到顶部。然而，非常小的液滴（如雾流）很难靠重力分离。这些液滴可以聚结成大液滴，然后再靠重力分离。

2. 烟囱用来平衡下部和气体区的压力。降液管的出口或铺散器位于油水界面处。当油在这里上升时，任何油相中的自由水都将分离出来。水滴流向与油流相反。同理，当水向下流时，束缚在水相中的油滴逆流向上。

4.3　多级分离

🔲 导语

　　应用多级分离可使气液两相更好地分离，以便使烃类液体回收率最大化。根据气油比和井流压力，分离级数通常介于两级至四级。分级分离的主要目的是使经过最后分离的最终相（气体和液体）达到最大程度的稳定，即最后的气相和液相中都不会再有大量的液相或气相逸出。

🔲 课文

多级分离

　　为了使气液两相更好地分离，以便使烃类液体回收率最大化，要采取多级降压分离，即让井流物通过两个或两个以上的系列分离器。操作压力依次降低，第一个分离器压力最高，最后一个分离器压力最低。实际上，分离级数通常介于两级至四级，这取决于气油比和井流压力。两级分离通常用于低气油比和低井流压力；三级分离用于中高气油比和中等井流压力；四级分离适用于高气油比和高井流压力。据记录，三级分离通常经济性最优，和两级分离相比，其液体回收率高出 2% ~ 12%，在某些情况下甚至能达到 25% 以上。为了回收在中至低压下运行的分离器所产生的气体组分，有必要将其再压缩到高压分离器所具有的压力。不过对于伴生气体，再压缩成本太高，因此从低压分离器产生的气体可以用火炬烧掉。

　　应当指出的是，分级分离的主要目的是使经过最后分离器后的最终相（气体和液体）达到最大程度的稳定，这意味着最后的气相和液相中分别不会再有数量可观的液相或气相逸出。

离心分离器

在离心分离器或旋风分离器中,当液滴进入一个圆柱形分离器时,离心力对雾滴的作用力是重力的好几倍。在低速大离心机中,离心力可达重力的 5 倍;而在小的高压离心机中,离心力可达重力的 2000 倍。一般来说,离心分离机可以用来分离直径大于 $100\,\mu m$ 的液滴,而规格适当的离心分离机对尺寸小至 $10\,\mu m$ 的液滴也有一定的分离效率。

离心分离器分离携带颗粒较多的气流也是相当有效的。圆柱形气液旋风式分离器尺寸紧凑,占地面积小,质量小,具有降低工业成本的应用潜力,尤其适于海上应用。同时圆柱形气液旋风式分离器明显降低了碳氢化合物的总量,这一点对于环保和安全十分重要。圆柱形气液旋风式分离器主要用于大批气液分离,可根据预期性能设计为不同级别。

扭转式超声分离器

扭转式超声分离器综合了一系列物理处理过程,包括膨胀、旋流式气液分离以及再压缩,在紧凑的管状装置中凝结和分离天然气中的水和重烃类组分。在超音速下实现凝结和分离,是该分离器能够降低固定资本和操作成本的关键。扭转式超声分离器内的介质停留时间仅仅只有几毫秒,来不及形成水合物,就不需要加入水合物化学抑制剂了。不使用相关化学物质的再生系统,就可以避免苯、甲苯、二甲苯等有害物质排放到环境中,或者说不再需要支付化学物质回收系统的费用。这种简单可靠的固定装置无旋转部件,不用加入化学物质,非常适合在恶劣环境或近海环境中广泛进行无人操作。此外,简易、小质量的气旋式系统设计,使其能在空间和质量都受限的平台应用。

段塞流捕集器

段塞流捕集器用于近海管道末端来捕获管道中的液体大段塞,并暂时储存这些段塞,然后使其以适当流速进入下游设备和处理设施。段塞流捕集器可能是一个容器罐或管构件。相同容量的管型段塞流捕集器往往比罐式的便宜,因为可采用管壁较薄的小直径管。多管型段塞流捕集器可重叠布置,这样可以通过放置更多平行的管道增加容量。管式段塞流捕集器一般配置包括以下部分:

(1)指形管有双斜面和三个分区:气液分离区、中间区、储存区。

(2)天然气立管在分离区和中间区的过渡带处与每个指形管相连。

(3)天然气均气管布置在每个指形管上。这些线路都设在段塞储存段。

(4)液体汇管箱从每个指形管内收集液体。汇管箱不能倾斜,垂直于指形管安装。

需注意的是,尽管在指形管内可能把凝析油和水直接分开,仍假定所有的液体(凝析油和水)收集后送到三相分离器。当段塞流捕集器在其自身内进行凝析油水分离时,要让最大的凝析油段塞和最大的水段塞保持分离,以确保连续的液位控制。

气体和液体两相的分离是在指形管的第一段实现的。该段管子的长度可以促进层流流动并实现初步分离。在理想的情况下,600 μm 及以下的液滴可从进入天然气立管的气体中分离出来,该立管安装在本段的末尾。中间段的长度是最小的,保证液体段塞捕集器充满的时候,天然气立管下没有液面,即储存区完全充满。本段在天然气立管及储存段之间可以有高度变化,以便明确区分气液两相。

储存段的长度要确保最大段塞储集体积,不能让液体进入天然气出口。正常运行期间,正常液面保持在立管的顶部上下,液体从每个指形管流入主液体收集汇管箱,这相当于凝析稳定装置以最大容量运行大约 5min。

段塞流捕集器的设计取决于几个因素,其中最重要的是清管作业和流量改变。清管可以使捕集器尺寸不用过大,因为频繁清管可以减少管道中液流的累积,最大段塞尺寸就可以减小在一定范围内。但段塞流捕集器尺寸大小的选择,应在频繁清管作业成本增加和小型段塞流捕集器的投资减少之间权衡。

🔳 Key to Exercises

I.1. The main objective of stage separation is to provide maximum stabilization to the resultant phases (gas and liquid) leaving the final separator, which means that the considerable amounts of gas or liquid will not evolve from the final liquid and gas phases, respectively.

2. The compact dimensions, smaller footprint, and lower weight of the GLCC have a potential for cost savings to the industry, especially in offshore applications. Also, the GLCC reduces the inventory of hydrocarbons significantly, which is critical to environmental and safety considerations. The GLCC separator, used mainly for bulk gas/liquid separation, can be designed for various levels of expected performance.

3. The general configuration consists of the following parts.

(1) Fingers with dual slope and three distinct sections: gas/liquid separation, intermediate, and storage sections.

(2) Gas risers connected to each finger at the transition zone between sep-

aration and intermediate sections.

(3) Gas equalization lines located on each finger. These lines are located within the slug storage section.

(4) Liquid header collecting liquid from each finger. This header will not be sloped and is configured perpendicular to the fingers.

4. The Twister supersonic separator is a unique combination of known physical processes, combining expansion, cyclonic gas/liquid separation, and recompression process steps in a compact, tubular device to condense and separate water and heavy hydrocarbons from natural gas. Condensation and separation at supersonic velocity are key to achieving step – change reductions in both capital and operating costs.

5. Slug catcher design is dependent on several factors, of which the most important are pigging operation and changes in flow rates.

Ⅱ. 1. B 2. C 3. D 4. A 5. B 6. A 7. A 8. D 9. C 10. B

Ⅲ. 1. C 2. D 3. A 4. D 5. A 6. B 7. D 8. C 9. C 10. D

Ⅳ. 1. 一般来说,离心分离机可以用来分离直径大于 $100\mu m$ 的液滴,而规格适当的离心分离机对尺寸小至 $10\mu m$ 的液滴也有一定的分离效率。

2. 圆柱形气液旋风式分离器尺寸紧凑,占地面积小,质量小,具有降低工业成本的应用潜力,尤其适于海上应用。

3. 当段塞流捕集器在其自身内进行凝析油水分离时,要让最大的凝析油段塞和最大的水段塞保持分离,以确保连续的液位控制。

4. 正常运行期间,正常液面保持在立管的顶部上下,液体从每个指形管流入主液体收集汇管箱,这相当于凝析稳定装置以最大容量运行大约5min。

5. 清管可以使捕集器尺寸不用过大,因为频繁清管可以减少管道中液流的累积,最大段塞尺寸就可以减小在一定范围内。

Ⅴ. 1. 为了使气液两相更好地分离,以便使烃类液体回收率最大化,要采取多级降压分离,即让井流物通过两个或两个以上的系列分离器。操作压力依次降低,第一个分离器压力最高,最后一个分离器压力最低。实际上,分离级数通常介于两级至四级,这取决于气油比和井流压力。两级分离通常用于低气油比和低井流压力;三级分离用于中高气油比和中等井流压力;四级分离适用于高气油比和高井流压力。

2. 扭转式超声分离器内的介质停留时间仅仅只有几毫秒,来不及形成水合物,就不需要加入水合物化学抑制剂了。不使用相关化学物质的再生系统,就可以避免苯、甲苯、二甲苯等有害物质排放到环境中,或者说不再需要支付化学物质回收系统的费用。这种简单可靠的固定装置无旋转部件，不用加入化学物质，非常适合在恶劣环境或近海环境中广泛进行无人操作。

4.4 天然气凝析液回收

导语

天然气处理一般是将天然气中较重的液态烃组分除去。提取天然气凝析液不仅是为了控制天然气露点，而且 NGL 作为独立的市场产品比其在天然气中价值更高，可以获得更多收益。因此，为使天然气生产效益最大化，单独回收 NGL 具有经济意义。然而，无论利润如何，天然气加工处理必须符合安全运输和安全燃烧的标准。

课文

天然气中较重的液态烃组分通常称为天然气凝析液（NGL），包括乙烷、丙烷、丁烷和天然汽油（凝析油）。天然气较轻的液态烃组分可以作为燃料或者原料销售给炼油厂或石油化工厂；而较重部分可以用作调和汽油。NGL 作为天然气液和作为燃料的销售价格差别很大，这个差价通常称为"缩水价值"，常常决定着天然气处理厂回收程度的高低。当然，无论经济效益如何，天然气加工处理必须符合安全运输和安全燃烧的标准，因此回收利润率并不是决定天然气液提取程度的唯一标准。天然气液回收一般在比较集中的处理厂进行，回收的天然气液经过处理符合销售标准后，再进入天然气液输送设施。

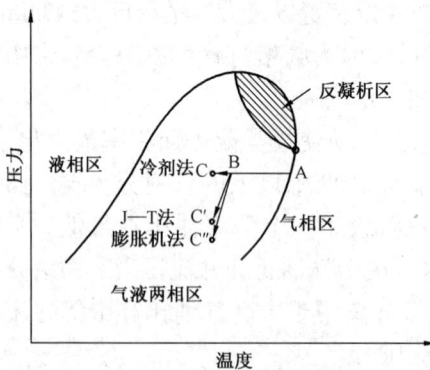

图 4.4.1 各种 NGL 提取技术热力学路径

NGL 回收方法

图 4.4.1 表明了不同温度和压力下天然气的相态特征。很明显，相图中在相包络线内除反凝析以外的区域，随着温度的降低，天然气液会逐渐析出。反凝析现象在天然气液生产中起着重要的作用。有些工厂生产时初始压力高于临界点压力，当温度降低到临界凝析温度以下时，天然气液就会再次气化。因此，确定温度正处于相包络线上的哪点至关重要。

制冷法

制冷法就是在不同温度下冷凝或冷却气体、蒸汽或者液体的工艺。机械制冷法是回收 NGL 最简单和最直接的方法。

贫油吸附法

NGL 回收的吸附方法与天然气脱水使用的吸附方法非常类似，主要的不同是，在 NGL 吸附中使用吸附油，而天然气脱水时使用二乙醇。这种吸附油对 NGL 具有亲和力，这与乙二醇对水有亲和力是一样的。

贫油吸附法是最古老的、效率最低的 NGL 回收方法。值得注意的是，油吸附装置不能有效地回收乙烷和丙烷，需要循环使用大量的吸附油，要求额外维护，并且消耗过多的燃料，因此现在已经很少使用和制造贫油吸附装置了。由于贫油吸附装置造价昂贵，操作更复杂，并且贫油会随时间变质，因此难以预测从天然气中除去液体的效率。

固相层吸收法

这种方法使用能够吸附天然气中重烃的吸附剂。吸附剂可以是硅胶或活性炭，有重烃时不能使用活性氧化铝，因为此时重烃会污染吸附剂。

这种工艺适合浓度相对低的重烃。如果气体处于接近临界凝析压力的高压，这种工艺也非常适用。在这种情况下，制冷过程无效，吸附分离成为达到所需技术标准唯一的方法。

薄膜分离法

冷冻和低温装置一直以来用于 NGL 回收，这些装置投资成本和操作成本都很高，此外因为有大量的旋转部件，操作起来很复杂。

薄膜分离法从天然气中去除和回收重烃简单且成本低。分离过程是基于高流量的薄膜，这种薄膜可选择性地渗透比甲烷重的烃。这些烃穿过薄膜，通过再压缩和冷凝，作为液体回收。除去了部分重烃的薄膜剩余流体送去销售。

薄膜系统用途很多，可以设计处理多种进料。因为质量小、结构紧凑，薄膜系统非常适合海上应用。

天然气液分馏

天然气液回收装置的底部流体可以作为混合产品出售，这在当地需求不足的偏僻小型工厂很常见，混合产品通过卡车、铁路、驳船或管线送到指定地点进一步处理。更经济的方法是把液体分馏成不同组分，这些组分作为单纯产品更有市场价值，然而由于天然气和 NGL 的相对价格波动，从天然气中提取 NGL 的相对动力就会有所变化，且从天然气中提取 NGL 的程度可自由确定：安全因素决定着最低的提取程度，技术和 NGL 相对市场价值之间的平衡决定着最高的提取程度。

将 NGL 气流分离成不同组分的工艺称为分馏。在分馏装置中,液体分离成有商品价值的产品,然后通过油轮和油罐车进入市场。加热混合的 NGL 流体,然后通过一系列的蒸馏塔进行分馏。分馏就是利用了不同 NGL 产品有不同沸点这一特性,随着 NGL 流体温度的增加,最轻的 NGL 产品(沸点最低)在塔顶汽化,然后冷凝成纯净的液体流到储存罐,塔底较重的液体混合物流到第二个塔,重复分馏过程,分馏出不同产品并储存起来。这种工艺一直重复到 NGL 分离出各种组分。

汽油和液化石油气加工

天然汽油(凝析液)和液化石油气(LPG)经常受酸性化合物污染,特别严重的是硫化氢、硫醇和单质硫,含硫化氢的天然汽油有难闻的气味且具腐蚀性,硫醇也使汽油难闻,单质硫使汽油具有腐蚀性。

作为进料时,LPG 中的硫化氢会形成游离硫或硫醇。如果硫醇含量很大,会使液化石油气十分难闻,且燃烧的产物也具有难闻的气味。二氧化碳含量过高会增加 LPG 压力,降低发热量。羰基硫和二硫化碳虽然没有腐蚀性,但会在自由水中缓慢水解成硫化氢,使产品具有腐蚀性。

液—液接触法使用碱液、醇胺水溶液或固体氢氧化钾从 LPG 和汽油中除去硫化氢和二氧化碳。当硫化氢和二氧化碳组分含量很低时,简单的碱洗既经济又有效。但是随着污染物含量增加,供应和处理苛性碱成本高,这种方法就不再实用了。胺处理法是非常不错的备选方案,特别是现场有胺处理装置时更是如此。

Key to Exercises

Ⅰ.1. Recovery of NGL components in gas not only may be required for hydrocarbon dew point control in a natural gas stream, but also yields a source of revenue, as NGLs normally have significantly greater value as separate marketable products than as part of the natural gas stream.

2. NGL recovery processes include refrigeration processes, lean oil absorption, solid bed adsorption and membrane separation process.

3. The membrane separation process offers a simple and low – cost solution for removal and recovery of heavy hydrocarbons from natural gas. The separation process is based on a high – flux membrane that selectively permeates heavy hydrocarbons compared to methane. These hydrocarbons permeate the membrane and are recovered as a liquid after recompression and condensation. The

residue stream from the membrane is partially depleted of heavy hydrocarbons and is then sent to a sales gas stream.

4. NGLs are fractionated by heating mixed NGL streams and passing them through a series of distillation towers. Fractionation takes advantage of the differing boiling points of the various NGL products. As the temperature of the NGL stream is increased, the lightest NGL product boils off the top of the tower as a gas where it is then condensed into a purity liquid that is routed to storage. The heavier liquid mixture at the bottom of the first tower is routed to the second tower where the process is repeated and a different NGL product is separated and stored. This process is repeated until the NGLs have been separated into their components.

5. H_2S and CO_2 can be removed from LPG and gasoline by liquid – liquid contacting processes using a caustic solution, aqueous alkanolamines, or solid KOH. When quantities of H_2S and CO_2 components are small, a simple caustic wash is both effective and economical. However, as the quantity of contaminants rises, the caustic supply and disposal costs render this approach impractical. Amine treating is a very attractive alternative, especially when there is already an amine gas treating unit on site.

Ⅱ. 1. B 2. C 3. A 4. D 5. B 6. A 7. C 8. D 9. B 10. A

Ⅲ. 1. B 2. B 3. D 4. C 5. A 6. C 7. A 8. D 9. C 10. B

Ⅳ. 1. NGL 作为天然气液和作为燃料的销售价格差别很大,这个差价通常称为"缩水价值",常常决定着天然气处理厂回收程度的高低。

2. 当然,无论经济效益如何,天然气加工处理必须符合安全运输和安全燃烧的标准,因此回收利润率并不是决定天然气液提取程度的唯一标准。

3. 这种方法使用能够吸附天然气中重烃的吸附剂。

4. 塔底较重的液体混合物流到第二个塔,重复分馏过程,分馏出不同产品并储存起来。

5. 天然气液回收一般在比较集中的处理厂进行,回收的天然气液经过处理符合销售标准后,再进入天然气液输送设施。

Ⅴ. 1. 贫油吸附法是最古老的、效率最低的 NGL 回收方法。值得注意的是,油吸附装置不能有效地回收乙烷和丙烷,需要循环使用大量的吸附油,要求额外维护,并且消耗过多的燃料,因此现在已经很少使用和制造贫油吸附装置了。由于贫油吸附装置造价昂贵,操作更复杂,并且贫油会随时间变质,因此难以预测从天然气中除去液体的效率。

2. 天然气液回收装置的底部流体可以作为混合产品出售,这在当地需求不足的偏僻小型工厂很常见,混合产品通过卡车、铁路、驳船或管线送到指定地点进一步处理。更经济的方法是把液体分馏成不同组分,这些组分作为单纯产品更有市场价值,然而由于天然气和 NGL 的相对价格波动,从天然气中提取 NGL 的相对动力就会有所变化,且从天然气中提取 NGL 的程度可自由确定:安全因素决定着最低的提取程度,技术和 NGL 相对市场价值之间的平衡决定着最高的提取程度。

第5章 天然气酸性气体处理

5.1 酸性气体处理(1)

导语

本章中的"酸性气体"一词是指含有大量的硫化氢(H_2S)、二氧化碳（CO_2）或类似酸性气体的天然气。"酸性气体"和"酸气"常常错误地当作同义词。严格来说，酸气是含有大量硫化氢的气体；不含硫化氢的天然气为甜气；而酸性气体中含有大量酸性气体，如二氧化碳或硫化氢。因此，二氧化碳本身是一种酸性气体而不是酸气。

课文

自然界中以烃类形式存在的天然气中含有大量的酸性气体，如硫化氢和二氧化碳。含有硫化氢或二氧化碳的天然气是酸气，不含硫化氢的天然气为甜气。一旦有水，硫化氢和二氧化碳就会产生腐蚀性(形成一种酸性水溶液)，而且硫化氢有毒，二氧化碳没有热值，因此，待售天然气的硫化氢含量须不高于 $5mL/m^3$，并且热值不低于 $920 \sim 980$ Btu/ft^3。实际的标准要视天然气的用途以及用气国家和合约限制而定。然而，由于天然气组分繁多，且含有两种酸性气体，因此各种酸性气体脱除法也各异，其选择取决于最终产品的要求。

天然气处理过程中有许多变量，某一方法的适用范围很难定义。

天然气处理必须考虑诸多因素：(1)气体中杂质的类型和含量；(2)净化的质量要求；(3)需要选择的酸性气体脱除；(4)待处理气体的温度、压力、体积和组成；(5)气体中二氧化碳与硫化氢的比率；(6)出于经济成本或环境问题的考虑回收硫的必要性。

除了硫化氢和二氧化碳，天然气还可能含有其他杂质，例如硫醇和羰基硫化物。这些杂质的存在导致脱去大量酸性气体后却并未能将酸性气体的浓度降至足够低，减弱了净化过程的效果。然而，有一些不是按照去除(或不能去除)大量酸性气体设计的流程，能在气体中的酸性气体处于中低浓度时将酸性气体杂质降到很低的程度。

工艺选择是指选择一种能更好地脱去某种酸性气体组分的方法。例如，一些工艺可以脱除硫化氢和二氧化碳，而一些工艺只能脱除硫化氢。考虑工艺选择十分重要,比如硫化氢脱除法和二氧化碳脱除法相比,二氧化碳脱除法能保证产品中这些组分含量最低,因此就要考虑选择脱除气体中的二氧化碳而不是硫化氢。

酸性气体脱除方法

天然气净化的方法从简单的一次过洗作业发展到复杂的多步循环系统。大多数情况下,用于清除杂质的材料需要回收,甚至需要回收杂质或者其转变物,这使工艺过程更加复杂。

常用于酸性气体脱除的方法有两种:吸附和吸收。吸附是一种物理化学现象,通过固体或液体将杂质气体吸收在其表面来清除杂质。一般地,活性炭可作为吸附介质,并在解吸后可以再次使用。吸附量与固体的表面积成正比,因此,吸附剂通常是具有较大比表面的颗粒状固体。吸附气体可通过热空气或热气流解吸后再回收或热解。吸附剂广泛用于煅烧前提高气体浓度。吸附也用来去除气体中的异味。吸附系统的使用有很多限制条件,但最重要的是减少气流中固体颗粒和液体(例如水蒸气)含量,因为它们可能会堵住吸附剂的表面从而大大降低吸附效率。

吸收不同于吸附,它不是物理化学表面现象,吸收气最终完全分散于吸收剂(液体)中。该过程仅依赖于溶解,在液相中会发生化学反应(化学吸收)。常用的吸收介质有水、醇胺水溶液、碳酸钠溶液和非挥发性的液态烃,根据要处理的气体选择对应的吸收剂。通常用的吸收塔是板式塔或填料塔。

吸收作用通过溶解作用(物理现象)或化学反应(化学现象)来实现。化学吸附过程是将二氧化硫吸附到炭表面,而后氧化(由烟道气中的氧气氧化),再吸收水分生成硫酸浸入到吸附剂中。

目前使用的脱酸性气体方法包括酸性气体和固体氧化物(如氧化铁)的化学反应,或有选择性地将污染物吸收到液体(如乙醇胺)中,其中液体与气体逆向流动。然后吸收剂解吸气体成分并再循环至吸收器中。实际上,过程设计是多样的,并可能使用多级吸收器和多级再生器。

液体吸收方法[温度一般低于 $50℃$($120℉$)]分为物理溶剂法和化学溶剂法。物理溶剂法常用有机溶剂,一般在高压和较低温度下进行。化学溶剂法主要是用碱溶液如醇胺类或碳酸盐类来吸收酸性气体。通过降压或升温进行再生(解吸),使酸性气体从溶剂中分离出来。

醇胺法除酸性气体过程中,醇胺和酸性气体发生化学反应可以释放出大量热量,足够补偿吸收的热量。胺的衍生物,如乙醇胺(一乙醇胺)、二乙醇胺、三乙醇胺、甲基二乙醇胺、二异丙醇胺和二甘醇胺,已广泛应用于商业脱酸性气体。

□ **Key to Exercises**

Ⅰ. 1. There are many variables in treating natural gas. Several factors must be considered: (1) types and concentrations of contaminants in the gas, (2) the degree of contaminant removal desired, (3) the selectivity of acid gas removal required, (4) the temperature, pressure, volume, and composition of the gas to be processed, (5) the carbon dioxide – hydrogen sulfide ratio in the gas, and (6) the desirability of sulfur recovery due to process economics or environmental issues.

2. Process selectivity indicates the preference with which the process removes one acid gas component relative to (or in preference to) another. For example, some processes remove both hydrogen sulfide and carbon dioxide; other processes are designed to remove hydrogen sulfide only.

3. The two general processes used for acid gas removal are: adsorption and absorption. Adsorption is a physical – chemical phenomenon in which the gas is concentrated on the surface of a solid or liquid to remove impurities. Absorption differs from adsorption in that it is not a physical – chemical surface phenomenon, but an approach in which the absorbed gas is ultimately distributed throughout the absorbent (liquid). The process depends only on physical solubility and may include chemical reactions in the liquid phase (chemisorption).

4. As currently practiced, acid gas removal processes involve the chemical reaction of the acid gases with a solid oxide (such as iron oxide) or selective absorption of the contaminants into a liquid (such as ethanolamine) that is passed countercurrent to the gas.

5. Amine derivatives such as ethanolamine (monoethanolamine), diethanolamine, triethanolamine, methyldiethanolamine, diisopropanolamine, and diglycolamine have been used in commercial applications.

Ⅱ. 1. A 2. A 3. D 4. C 5. B 6. C 7. D 8. B 9. B 10. A

Ⅲ. 1. B 2. D 3. C 4. D 5. C 6. A 7. B 8. A 9. D 10. D

Ⅳ. 1. 这些杂质的存在导致脱去大量酸性气体后却并未能将酸性气体的浓度降至足够低,减弱了净化过程的效果。

2. 大多数情况下,用于清除杂质的材料需要回收,甚至需要回收杂质或者其转变物,这使工艺过程更加复杂。

3. 化学吸附过程是将二氧化硫吸附到炭表面,而后氧化(由烟道气中的氧气氧化),再吸收水分生成硫酸浸入到吸附剂中。

4. 目前使用的脱酸性气体方法包括酸性气体和固体氧化物(如氧化铁)的化学反应,或有选择性地将污染物吸收到液体(如乙醇胺)中,其中液体与气体逆向流动。

5. 醇胺法除酸性气体过程中,醇胺和酸性气体发生化学反应可以释放出大量的热,足够补偿吸收的热量。

Ⅴ.1. 自然界中以烃类形式存在的天然气中含有大量的酸性气体,如硫化氢和二氧化碳。含有硫化氢或二氧化碳的天然气是酸性的,不含硫化氢的天然气为甜气。一旦有水,硫化氢和二氧化碳就会产生腐蚀性,而且硫化氢有毒,二氧化碳没有热值,因此,待售天然气的硫化氢含量须不高于 $5mL/m^3$,并且热值不低于 $920\sim980Btu/ft^3$。实际标准要视天然气的用途以及用气国家和合约限制而定。然而,由于天然气组分繁多,且含有两种酸性气体,因此各种酸性气体脱除法也各异,其选择取决于最终产品的要求。

2. 吸附是一种物理化学现象,通过固体或液体将杂质气体吸收在其表面来清除杂质。一般地,活性炭可作为吸附介质,并在解吸后可以再次使用。吸附量与固体的表面积成正比,因此,吸附剂通常是具有较大比表面的颗粒状固体。吸附气体可通过热空气或热气流解吸后再回收或热解。吸收不同于吸附,它不是物理化学表面现象,吸收气最终完全分散于吸收剂(液体)中。该过程仅依赖于溶解,在液相中会发生化学反应(化学吸收)。

5.2　酸性气体处理(2)

🔲 导语

含有硫化氢和/或二氧化碳的原料天然气必须经过处理以去除杂质,直至符合标准才能使用。天然气处理过程中有许多变量,由于诸多制约因素,某一方法的适用范围很难定义。除了上一节讲的两种常用的酸性气体脱除方法:吸附和吸收,本文还将涉及间歇处理法、醇胺法、碱洗和水洗法、甲醇法。

🔲 课文

间歇处理法

最常见的酸性气体脱除方法是间歇处理法,其中一般涉及酸性气体与清洗剂发生化学反应。清洗剂一般为金属氧化物。处理过程不仅仅包括物理过程,例如吸附等去除酸性气体的物理现象。间歇处理的各个流程没有特殊的技术要求,在循环终端对化学处理剂进行更换或再生利用。间歇处理流程只能脱除少量的硫,例如天然气流速低和/或硫化氢的浓度低。

该方法可通过使原料气与固体介质反应,清除气流中硫化氢和有机硫化物(硫醇)。固体介质一般是不可再生的,即使有些是部分可再生的,在各个再生循环中也逐步失去活性。大多数干式吸附工艺是金属氧化物与硫化氢反应生成金属硫化物。再生反应中,金属硫化物与氧气反应得到硫和再生的金属氧化物。在干式吸附脱酸工艺中应用的主要金属氧化物是氧化铁和氧化锌。

醇胺法

化学吸收法利用醇胺水溶液处理含有二氧化碳和硫化氢的天然气流。但是,需要根据待处理天然气的组成来选择不同的胺,以满足不同规格产出气要求。根据有机基团对中心氮原子的取代度,可以将胺分为伯胺、仲胺、叔胺。伯胺可以直接与 H_2S、CO_2 以及羰基硫(COS)反应,其代表是一乙醇胺(MEA)和专用二乙二醇胺(DGA)。仲胺可直接与 H_2S 和 CO_2 反应,并可直接与部分羰基硫反应,最重要的仲胺是二乙醇胺(DEA)。叔胺则直接与 H_2S 反应,间接与 CO_2 反应,并间接与少量的羰基硫反应。甲基二乙醇胺(MDEA)和活性甲基二乙醇胺是最常见的叔胺品种。

现在广泛使用乙醇胺和磷酸钾进行天然气除酸。使用乙醇胺的处理方法叫做乙醇胺法,可以除去液态烃、天然气和炼厂气中的酸性气体(H_2S 和 CO_2)。根据需要,常会配制特殊溶液,如不同胺的混合溶液、如加入环丁砜和六氢吡嗪等物理溶剂的胺溶液,还有被酸(如磷酸)部分中和的胺溶液。

一乙醇胺(MEA)比较稳定,如果不受其他化学物质影响,在温度达到其沸点的情况下也不会降解或分解。二乙醇胺(DEA)的碱性和腐蚀性比一乙醇胺弱,所以二乙醇胺系统不会像一乙醇胺那样受到腐蚀问题的困扰,但确实会和硫化氢及二氧化碳反应。二乙醇胺(DEA)也能部分去除羰基硫和二硫化碳,因为二乙醇胺可再生,与羰基硫和二硫化碳反应化合后不会损失。

不同胺的一个重要区别是其对硫化氢的选择性。某些胺类可以将硫化氢脱除到符合标准要求,并允许一定量的二氧化碳通过,不像一乙醇胺和二乙醇胺那样既清除硫化氢又清除二氧化碳。

碱洗和水洗法

碱洗是一种为了控制排放、清除气流中酸性气体(如二氧化碳和硫化氢)的弱碱处理过程,其原理是碳酸钾吸收二氧化碳的速率随温度的升高而升高。实践证明,该方法在能发生逆向反应的温度条件下运行最好。

从结果来看,水洗与用碳酸钾洗涤效果相同,在压力逐渐下降的情况下,也能发生解吸作用。该吸收过程是纯粹的物理过程,而且吸收烃类的能力也较高,解吸时烃类和酸性气体同时被释放出来。

甲醇法

甲醇是天然气加工业中用途最广的溶剂之一。从历史上看,甲醇是第一个商用的有机溶剂,并已用于水合物抑制、脱水、天然气净化和液体回收。大多是在低温下使用甲醇,因为其他溶剂在低温时黏度都比较大,甚至容易凝固,而甲醇却不会。低温下操作,还可以避免甲醇最大的弱点——高挥发损耗。再者,甲醇容易生产,相对便宜,这些优点使甲醇应用更为广泛。

除了蒸气压力较高,甲醇的物理特性都优于其他溶剂。甲醇黏度低的优点在减小冷箱注入装置的压降和提高热传导方面有明显的作用。甲醇的表面张力比其他溶剂的小得多。表面张力过高容易使接触器里起泡,而甲醇不易起泡。然而,甲醇的主要缺点是蒸气压力高,比乙二醇和胺要高好几倍。为了减少甲醇损失,促进水和酸性气体的吸收反应,吸收器和分离器的温度一般不高于 -20 °F。

甲醇蒸气压力高可能导致溶剂损失较大,初看起来这是很大的缺点,但蒸气压较高也有明显的优势。尽管通常不考虑,但气体与溶剂不能充分混合也能带来大问题。由于蒸气压力高,甲醇在进入冷箱之前可以与气体充分混合,而乙二醇却要在冷箱里采用专门的喷嘴并设定各喷嘴位置来防止其凝结。溶剂带入下一个流程也是一个重要的问题。因为甲醇比乙二醇、胺及其他有机溶剂(包括贫油)更容易挥发,甲醇通常会在下游再生过程中排出。甲醇在汽提塔顶部的冷却器里分离出来并进一步提纯。可是如果乙二醇带入醇胺装置,就会溶解在溶液中,可能会降解并稀释醇胺溶液。

🔲 Key to Exercises

Ⅰ. 1. The most common type of process for acid gas removal is the batch-type process and may involve a chemical process in which the acid gas reacts chemically with the cleaning agent, usually a metal oxide.

2. Chemical absorption processes with aqueous alkanolamine solutions are used for treating gas streams containing hydrogen sulfide and carbon dioxide. However, depending on the composition and operating conditions of the feed gas, different amines can be selected to meet the product gas specification.

3. One key difference among the various specialty amines is selectivity to-

ward hydrogen sulfide. Instead of removing both hydrogen sulfide and carbon dioxide, as generic amines such as MEA and DEA do, some products readily remove hydrogen sulfide to specifications, but allow controlled amounts of carbon dioxide to slip through.

4. Carbonate washing is a mild alkali process for emission control by the removal of acid gases (such as carbon dioxide and hydrogen sulfide) from gas streams and uses the principle that the rate of absorption of carbon dioxide by potassium carbonate increases with temperature.

5. Historically, methanol has been used for hydrate inhibition, dehydration, gas sweetening, and liquids recovery. Most of these applications involve low temperature where the physical properties of methanol are advantageous compared with other solvents that exhibit high viscosity problems or even solids formation. Operation at low temperatures tends to suppress the most significant disadvantage of methanol, high solvent loss. Furthermore, methanol is relatively inexpensive and easy to produce, making the solvent a very attractive alternate for gas processing applications.

Ⅱ. 1. C 2. D 3. A 4. C 5. B 6. D 7. A 8. D 9. A 10. B

Ⅲ. 1. A 2. B 3. A 4. C 5. D 6. C 7. D 8. B 9. A 10. B

Ⅳ. 1. 再生反应中,金属硫化物与氧气反应得到硫和再生的金属氧化物。

2. 叔胺则直接与硫化氢反应,间接与二氧化碳反应,并间接与少量的羰基硫反应。

3. 使用乙醇胺的处理方法叫做乙醇胺法,可以除去液态烃、天然气和炼厂气中的酸性气体。

4. 碱洗是一种为了控制排放、清除气流中酸性气体的弱碱处理过程,其原理是碳酸钾吸收二氧化碳的速率随温度的升高而升高。

5. 为了减少甲醇损失,促进水和酸性气体的吸收反应,吸收器和分离器的温度一般不高于 $-20\ {}^{\circ}\mathrm{F}$。

Ⅴ. 1. 最常见的酸性气体脱除方法是间歇处理法,其中一般涉及酸性气体与清洗剂发生化学反应。清洗剂一般为金属氧化物。处理过程不仅仅包括物理过程,例如吸附等去除酸性气体的物理现象。间歇处理的各个流程没有特殊的技术要求,在循环终端对化学处理剂进行更换或再生利用。间歇处理流程只能脱除少量的硫,例如天然气流速低和/或硫化氢的浓度低。

2. 甲醇蒸气压力高可能导致溶剂损失较大,初看起来这是很大的缺点,但蒸气压较高也有明显的优势。尽管通常不考虑,但气体与溶剂不能充分混合也能带来大问题。由于蒸气压力高,甲醇在进入冷箱之前可以与气体充分混合,而乙二醇却要在冷箱里采用专门的喷嘴并设定各喷嘴位置来防止其凝结。

5.3 酸性气体处理(3)

🔲 导语

酸性气体处理装置中出来的副产品主要是硫化氢和/或二氧化碳。二氧化碳一般排入大气或回注地下;然而,环保条例限制 H_2S 的排放。硫回收指的是将硫化氢转化为单质硫,最常用的转化方法是克劳斯工艺,约90% ~ 95%的硫回收使用的都是该工艺。克劳斯工艺是 1883 年由科学家 Carl Friedrich Claus 申请专利的,后来成为了行业标准。

🔲 课文

其他处理方法

磷酸钾脱硫和乙醇胺法脱除液态烃和天然气中的酸性气体一样,处理溶液是磷酸钾(K_3PO_4)水溶液,该溶液在吸收塔和再生塔中循环,与乙醇胺法中的乙醇胺循环方式是一样的;溶液可以加热再生。

其他方法包括:碱处理法;热碳酸钾处理法,将天然气和炼厂气中的酸性气体含量从50%降至0.5%,处理装置与胺处理装置类似。

砷碱法也可以用来清除硫化氢和/或二氧化碳。去除硫化氢时,反应物是碳酸钠或碳酸钾与亚砷酸盐和砷酸盐的混合物;去除二氧化碳时,用被三氧化二砷或亚硒酸或亚碲酸活化的热碱金属碳酸盐溶液。

分子筛去除气体中硫化氢(及其他硫化物)有很强的选择性,且能持续保持较高的吸收率。分子筛法对脱水同样有效,所以此方法可以同时用于脱水、脱硫,然而天然气含水过高时需要在上游脱水。由于分子筛吸收烃组分,部分天然气会损失。在这个处理进程中,不饱和烃组分如烯烃、芳香化合物会由分子筛吸附。分子筛易"中毒"于乙二醇这样的化学制剂,需要在进行吸附之前完全净化气体。

还有一种膜分离技术可用来处理含有 C_{3+} 碳氢化合物和/或酸性气体的天然气。处理后的天然气在气田或处理厂可作为燃气动力设备的燃料,包括气田或处理厂的压缩机。该方法也可以用于生产天然气液(NGL)。

工艺选择

相对于其他工艺来讲,前面介绍的每一种处理工艺在某些特定的应用领域都有自己的优势。因此,在选择合适的工艺时,要考虑下面几个方面:

(1)关于含硫化合物的处理空气污染条例规定和/或尾气净化要求;

(2)酸气中杂质组分的类型和含量;

(3)残气的质量要求;

(4)酸性气体要求;

(5)酸气的压力、温度和净化气的输送压力、温度;

(6)原料气的体积;

(7)原料气中的烃类含量;

(8)酸性气体脱除的选择性;

(9)投资费用和生产成本;

(10)设备的许可使用费;

(11)液体产品的质量要求;

(12)有害副产品的处理。

通常可以根据原料气组分和操作条件来简化处理方法的选择。酸性气体分压很高(50 psi)时可以使用物理溶剂来处理。原料气中重烃含量较高时则不用物理溶剂处理。若酸性气体分压低且对净化气的质量要求不高,通常选用胺处理。工艺选择并不容易,因为要考虑众多因素。初步评估之后,还需要进行相关的研究。

通常,陆上井口处理90%以上应用的是间歇处理法和醇胺法工艺。醇胺法是首选,因为费用较低,不用化学制剂,平衡了较高的设备费用。确定工艺方法的关键是原料气的硫含量。若硫含量低于20lb/d,间歇处理法更经济;若硫含量超过100lb/d,醇胺法更合适。

硫回收工艺

酸性气体处理装置中出来的副产品主要是硫化氢和/或二氧化碳。二氧化碳一般排入大气,但有时用作 CO_2 回注。H_2S 通常烧掉,转化成 SO_2 排放。环保条例限制 H_2S 的排放。而且还有很多特殊规定,这些规定也会定期修改。任何情况下,环保条例严格限制 H_2S 外排量及再生循环过程中燃烧的量。

大多数硫回收工艺是通过氧化 H_2S 产生硫单质。一般是通过 H_2S 和 O_2 或 H_2S 与 SO_2 两个反应,这两个反应都能生成水和单质硫。这些工艺都是有许可的,并用专门的催化剂和溶剂。这些工艺可直接用在采出气上。当气体流量很大时,通常让采出气与化学或物理溶剂接触,对再生过程中产出的酸性气体直接采用转换工艺处理。

　　硫回收通常有两种方法：液相氧化还原法和克劳斯回收法。

　　液相氧化还原法是采用化学吸收方法选择性地去除酸气中的硫化氢，使用含铁或矾的稀溶液。该工艺可在 H_2S 含量较少的情况下使用，某些情况下能取代酸性气体脱除工艺。可以先在入口处用弱碱性液体洗除 H_2S，并将其氧化成硫单质。还原的催化剂可在氧化装置中与空气接触再生。硫单质根据工艺可采用溶液悬浮或沉降法脱除。

　　克劳斯回收法是应用最为广泛的从酸气中回收单质硫的方法。克劳斯装置用于回收醇胺法再生器中排除的高含硫气流中的硫。这种方法可以处理硫化氢最大含量为 15% 的气体，其化学过程是将硫化氢氧化成二氧化硫，然后催化硫化氢和二氧化硫反应生成单质硫。

Key to Exercises

Ⅰ 1. The process using potassium phosphate is used in the same way as the Girbotol process to remove acid gases from liquid hydrocarbons as well as from gas streams. Other processes include the Alkazid process, the hot potassium carbonate process, the Giammarco – Vetrocoke process, molecular sieves and membrane – based process.

2. In selection of the appropriate process, the following facts should be considered.

（1）Air pollution regulations regarding sulfur compound disposal and/or Tail Gas Clean Up (TGCU) requirements .

（2）Type and concentration of impurities in the sour gas.

（3）Specifications for the residue gas .

（4）Specifications for the acid gas.

（5）Temperature and pressure at which the sour gas is available and at which the sweet gas must be delivered.

（6）Volume of gas to be processed.

（7）Hydrocarbon composition of the gas.

（8）Selectivity required for acid gas removal.

（9）Capital cost and operating cost.

（10）Royalty cost for process.

（11）Liquid product specifications.

（12）Disposal of by – products considered hazardous chemicals.

3. The key determinant is the sulfur content of the feed gas. When the sulfur content is below 20 – pound sulfur per day, batch processes are more economical, and when the sulfur content is over 100 – pound sulfur per day amine solutions are preferred.

4. There are two common methods of sulfur recovery: liquid redox and Claus sulfur recovery processes.

5. The Claus sulfur recovery process is the most widely used technology for recovering elemental sulfur from sour gas. The Claus process is used to recover sulfur from the amine regenerator vent gas stream in plants where large quantities of sulfur are present.

Ⅱ. 1. D 2. B 3. B 4. C 5. A 6. D 7. A 8. C 9. B 10. D

Ⅲ. 1. D 2. B 3. A 4. D 5. B 6. B 7. A 8. C 9. D 10. C

Ⅳ. 1. 热碳酸钾处理法将天然气和炼厂气中的酸性气体含量从50%降至0.5%,处理装置与醇胺处理装置类似。

2. 去除硫化氢时,反应物是碳酸钠或碳酸钾与亚砷酸盐和砷酸盐的混合物;去除二氧化碳时,用被三氧化二砷或亚硒酸或亚碲酸活化的热碱金属碳酸盐溶液。

3. 当气体流量很大时,通常让采出气与化学或物理溶剂接触,对再生过程中产出的酸性气体直接采用转换工艺处理。

4. 若硫含量低于20lb/d,间歇处理更经济;若硫含量超过100lb/d,醇胺法更合适。

5. 液相氧化还原法是采用化学吸收方法选择性地去除酸气中的硫化氢,使用含铁或矾的稀溶液。

Ⅴ. 1. 分子筛去除气体中硫化氢有很强的选择性,且能持续保持较高的吸收率。分子筛法对脱水同样有效,所以此方法可以同时用于脱水、脱硫,然而天然气含水过高时需要在上游脱水。由于分子筛吸收烃组分,部分天然气会损失。在这个处理进程中,不饱和烃组分如烯烃、芳香化合物会由分子筛吸附。分子筛易“中毒”于乙二醇这样的化学制剂,需要在进行吸附之前完全净化气体。

2. 通常可以根据原料气组分和操作条件来简化处理方法的选择。酸性气体分压很高时可以使用物理溶剂来处理。原料气中重烃含量较高时则不用物理溶剂处理。若酸性气体分压低且对净化气的质量要求不高,通常选用胺处理。工艺选择并不容易,因为要考虑众多因素。初步评估之后,还需要进行相关的研究。

第6章　天然气湿气输送

6.1　湿气输送

▫ 导语

　　海上气田空间受限,所有产出物经过简单处理后即输入管道,在管道内呈现多相流动状态,其中的流体可能是凝析油、水和天然气的三相混合物。多相流体可以用单一管道取代多管道分别输送,缩减运输成本。因此,单一管道就需要优化设计,设计者必须保证集输系统在设计的使用期限内必须安全、高效、可靠地运转。

▫ 课文

　　天然气产地与其市场通常不在同一地区,如世界上很多的海上气田,不可能在海上完成交易。为了满足市场需求,必须对天然气进行收集、处理和运输。很多时候,这些收集起来的天然气(一般含其他烃类较多,所以统称"湿气")会在不同管径的管道中经过相当长距离的运输才能到达市场。天然气管道的长度从数百英尺到数百英里不等,而且,沿途地区地势高低起伏,温度变化。天然气组分繁多,在管道运输中会因温度和压力的变化而发生相态变化,在管道中出现凝析液。此时,天然气在管道内的流动为两相流动(天然气/凝析油)。海上气田空间受限,所有产出物经过简单处理后即输入管道,在管道内呈现多相流动状态,因此,对于此类管道,优化设计显得尤为重要。这些管线在大洋底部处于水平或接近于水平的状态,其中的流体可能是凝析油、水(储层中固有)和天然气的三相混合物。

混相输送

　　混相输送技术对于海上边际油气田开发越来越重要,海上油气开发要利用现有的基础设施经济、有效地输送未处理的井筒流体,实现收益率最大化和固定资产投资及操作成本最小化。实际上,用不同管线输送并使用不同设备接收不同相态的流体,所需资金和空间都要多,可以用单一管道混相输送取而代之,大大缩减运输成本。混相输送技术减少了运输/装卸流体的管线数量,同时也减少了保养这些管线所需的工作,节省了水气相态分离和回注所需的支出以及管理费用。

设计者在对多相流集输系统进行热力和水力设计时面临着诸多难题，这些难题都与多相流动有关，多相流显著地提高了设计要求。所有管线设计者的目标都是确保流动安全，也就是说，集输系统在设计的使用期限内必须安全、高效、可靠地运转。如果做不到这点，就可能会导致严重的经济损失，特别是对于海上采气系统来说更是如此。流动保障措施涉及管线中可能遇到的所有问题，包括多相流动及与流体相关的影响因素，如天然气水合物的形成、管壁结蜡、沥青沉积在管壁上、腐蚀、冲蚀、结垢、乳化、起泡以及严重堵塞等。

天然气水合物

天然气水合物是一种呈笼状结构的冰状晶体，其中水分子构成主体结晶网格，网格中央包裹着其他小分子。最常见的小分子包括甲烷、乙烷、丙烷、异丁烷、正丁烷、氮气、二氧化碳、硫化氢等，其中天然形成的水合物中多为甲烷水合物。已经发现的天然气水合物的结构有很多种，最常见的两种为Ⅰ型和Ⅱ型。Ⅰ型天然气水合物网格中央包裹的小分子是甲烷、乙烷、硫化氢以及二氧化碳；Ⅱ型则由丙烷或者异丁烷之类的大分子形成金刚石型晶格。

虽然氮气是小分子，但形成的水合物也是Ⅱ型结构。此外，由于自由水的存在，温度和压力也会对水合物的结构产生影响，低温低压下的Ⅱ型水合物会随着温度和压力的升高逐渐转换成Ⅰ型。值得注意的是，正丁烷也可以形成水合物，但形成的水合物很不稳定。如果有甲烷或者氮气，正丁烷也可以形成稳定的水合物。由此可以推断，其他比正丁烷分子更大的正烷烃分子更加难以形成水合物。

水合物形成过程有很多影响因素，其中最主要的促使水合物形成的因素是：(1)合适的温度和压力；(2)气体温度等于或者低于该压力下天然气的水露点。需要注意的是，自由水虽然不是必要条件，但对水合物的形成起着极大的促进作用。影响天然气水合物形成的其他非必要因素包括紊流、结晶位置、结晶表面、晶体聚结、系统矿化度等。水合物的生成温度和压力跟天然气组成和水有关。在一定压力下，对任意组分的天然气来说，水合物生成温度是一定的，低于此温度才会有水合物形成，高于此温度则不会形成水合物。形成水合物的临界温度会随着压力增大而升高。一般说来，低温高压有利于水合物的形成。因此，许多天然气处理设备在关闭和重启时很容易形成水合物段塞。

天然气水合物无论是作为一种重要的碳氢化合物能源的来源，还是作为一种集输天然气的新方法，都具有可观的前景，但是水合物的形成却给生产造成了严重问题:水合物晶体可能在管壁沉降、逐渐累积甚至堵塞整个管道,最终导致生产中断。管道中压力逐渐降低会加速堵塞物的沉积,从而对生产设备造成相当严重的损害。此外,用较低的成本去除这种水合物堵塞物在技术上仍然比较困难。因此,如何预防天然气生产系统中形成水合物的问题多年来一直受到极大的关注。

从井口采出的多相流体通常温度适中而且压力比较高。当流体流入管线时,随着温度的降低,有可能在管线的某些地方形成水合物。因此,必须采取经济有效的措施预防水合物生成,以保证天然气集输管线正常运行。保持不利于水合物稳定的集输系统条件,就能防治水合物形成。例如,使流体温度高于水合物形成温度(包括合适的安全系数),或者使压力低于水合物生成压力,这两种方法基本上都是可行的。

🔲 Key to Exercises

Ⅰ. 1. Multiphase transportation technology has become more important for developing marginal fields, where the trend is to economically transport unprocessed well fluids via existing infrastructures, maximizing the rate of return and minimizing both capital expenditure and operational expenditure. By transporting multiphase well fluid in a single pipeline, separate pipelines and receiving facilities for separate phases, costing both money and space, are eliminated, which reduces capital expenditure.

2. The goal of any pipeline designer is to secure "flow assurance," i. e., the transmission system must operate in a safe, efficient, and reliable manner throughout the design life.

3. A gas hydrate is an ice – like crystalline solid called a clathrate, which occurs when water molecules form a cage – like structure around smaller guest molecules. The most common guest molecules are methane, ethane, propane, isobutane, normal butane, nitrogen, carbon dioxide, and hydrogen sulfide, of which methane occurs most abundantly in natural hydrates. Several different hydrate structures are known. The two most common are structure I and structure II. Type I forms with smaller gas molecules such as methane, ethane, hydrogen sulfide, and carbon dioxide, whereas structure II is a diamond lattice, formed by large molecules such as propane and isobutane.

4. Several different hydrate structures are known. The two most common are structure I and structure II. Type I forms with smaller gas molecules such as

methane, ethane, hydrogen sulfide, and carbon dioxide, whereas structure II is a diamond lattice, formed by large molecules such as propane and isobutane.

5. Control of hydrates relies on keeping the system conditions out of the region in which hydrates are stable. It may be possible to keep the fluid warmer than the hydrate formation temperature (with the inclusion of a suitable margin for safety) or operate at a pressure less than the hydrate formation pressure.

II. 1. B 2. A 3. D 4. A 5. B 6. C 7. A 8. A 9. D 10. C

III. 1. D 2. C 3. B 4. B 5. C 6. D 7. B 8. D 9. B 10. D

IV. 1. 混相输送技术对于海上边际油气田开发越来越重要,海上油气开发要利用现有的基础设施经济、有效地输送未处理的井筒流体,实现收益率最大化和固定资产投资及操作成本最小化。

2. 所有管线设计者的目标都是确保流动安全,也就是说,集输系统在设计的使用期限内必须安全、高效、可靠地运转。

3. 流动保障措施涉及管线中可能遇到的所有问题,包括多相流动及与流体相关的影响因素,如天然气水合物的形成、管壁结蜡、沥青沉积在管壁上、腐蚀、冲蚀、结垢、乳化、起泡以及严重堵塞等。

4. I型天然气水合物网格中央包裹的小分子是甲烷、乙烷、硫化氢以及二氧化碳;II型则由丙烷或者异丁烷之类的大分子形成金刚石型晶格。

5. 天然气水合物无论是作为一种重要的碳氢化合物能源的来源,还是作为一种集输天然气的新方法,都具有可观的前景,但是水合物的形成却给生产造成了严重问题:水合物晶体可能在管壁沉降、逐渐累积甚至堵塞整个管道,最终导致生产中断。

V. 1. 天然气的产地与其市场通常不在同一地区,如世界上的很多海上气田,不可能在海上完成交易。为了满足市场需求,必须对天然气进行收集、处理和运输。很多时候,这些收集起来的湿气会在不同管径的管道中经过相当长距离的运输才能到达市场。天然气管道的长度从数百英尺到数百英里不等,而且,沿途地区地势高低起伏,温度变化。天然气组分繁多,在管道运输中会因温度和压力的变化而发生相态变化,在管道中出现凝析液。

2. 水合物的生成温度和压力跟天然气组成和水有关。在一定压力下,对任意组分的天然气来说,水合物生成温度是一定的,低于此温度才会有水合物形成,高于此温度则不会形成水合物。形成水合物的临界温度会随着压力增大而升高。一般说来,低温高压有利于水合物的形成。因此,许多天然气处理设备在关闭和重启时很容易形成水合物段塞。

6.2 天然气水合物的预防技术

导语

多数天然气从井中采出或从伴生原油液流中分离出来时,都含有大量水蒸气。水蒸气必须从气流中脱除,因为从较高的气藏温度冷却至较低的地表温度时会凝析成液体,并形成水合物。水通常会加速腐蚀,并且集气系统内的固态水合物会压实,从而部分或完全阻塞采气管线。因此,水合物预防技术的研究引起了人们的重视,而海底天然气管道运输如何预防水合物则更具挑战性。

课文

由于天然气水合物形成的堵塞在集输管线中所处的位置不同,水合物的防治方法也有很多种,如安装加热炉或者降压输送,但是这些方法对于海底长距离高压集输管线可能就不适用了。海上油田天然气集输系统预防形成水合物的方法很少。一直以来,海上油田的防治措施是在天然气进入管线之前先用大型海上脱水装置对天然气进行脱水处理,但这种方法成本较高。通常,在现场有两种方法可以使用,即加热法和化学法。

加热法

加热法,即通过加热或者保温使混合流体的温度高于水合物生成温度。通常的做法是加入隔热层。这种方法可根据输送流体的种类、输送距离、平台面积大小等情况用于海底集输。保温系统的设计通常需要同时考虑材料成本(保温材料成本较高)、系统的可操作性、承受风险的能力三方面,在三者之间达到平衡。

给管线加热的方法也有许多种。对于双层套管或丛式管,最简单的方法是用外加水套炉,或者使用传导式或感应式伴热管线进行加热。为了保证传导系统的可靠性,隔热性能不好或者封闭条件下的长距离输送管线应该使用电阻加热方法。电阻加热装置的性能决定了它对生产系统的供热能力。

加热系统在控制流体温度时不会对环境造成危害,也不需要扩径以减小管线压力;而且,由于不用浪费时间去降压、清蜡、循环传热介质、解除水合物堵塞等,生产效率也提高了。但是,要想说服生产企业去安装一个这样的加热系统也很困难。

化学法

另一种预防水合物的方法是利用化学抑制剂。从井口加入化学抑制剂，使水合物生成的温度低于该管线的运转温度，从而预防水合物的生成。海底管线化学抑制剂注入系统的基础建设费用和运行费用都很高。尽管如此，运输未经处理的天然气应用最广泛的方法仍然是采用化学抑制剂，研发一种可替代的、低成本的、环保的水合物抑制剂，仍然是天然气生产工业的一个技术挑战。

抑制剂的种类

一般说来，天然气生产系统过去最常用的水合物的化学抑制剂是高浓度的甲醇、乙二醇或三甘醇。对于海上气田，设法增加采出水中盐含量（注入氯化钠、氯化钙、氯化钾之类的电解质溶液），也能降低水合物生成温度，然而仅仅靠这个方法是不够的。某些海上气田也会用甲醇或乙二醇与电解质溶液配制的混合抑制剂来抑制水合物的生成。

化学抑制剂的选择通常需要对很多因素进行比较，包括基础建设和运行费用、天然气的物理性质、安全因素、防腐以及天然气的脱水能力。但是，首要考虑的因素是化学药剂能否回收、再生、循环注入。然而，挥发损失是可以控制的，因此可以选用乙二醇作抑制剂。通常情况下，选择乙二醇作抑制剂成本很高，因为有甲醇损耗。由于甲醇的黏度和表面张力比较低，很容易在低温下（低于 -13 °F）从气体中分离出来，因此更受人们青睐。由于乙二醇这种抑制剂很昂贵，必须回收再生。无论是对于海上气田还是近海气田，乙二醇的回收再生装置都相当昂贵，而且需要占用很多空间。因此，需要研发一种新的、成本更低的、在低浓度下就能有效预防水合物形成的化学抑制剂。

现在，人们已经研发了两种新的、低剂量的、适用于海底天然气集输管线的化学抑制剂，不需要再注入乙二醇和增设乙二醇回收装置就可以输送更多的天然气。采用这种新型水合物抑制剂，不仅可以降低新型抑制剂的成本，而且还可以减小注入、输送以及储存设备的体积，因此可以显著降低成本。这些新型的水合物抑制剂称为"低剂量水合物抑制剂"，该项技术不用改变系统的热动力学条件就可以实施。

AAs 和其他类型的化学抑制剂不同，是一种表面活性剂，虽然不能抑制水合物晶体的形成，但是可以分散这些晶体小颗粒并防止其长大。这样，流体的黏度就会保持在较低的状态，形成的水合物就会随采出流体带走。相对来说，水合物的滞留时间对 AAs 的效果没有什么影响。另外，在很多恶劣条件下，AAs 比 KHIs 效果更好，因此，这类产品对于正在寻找廉价水合物抑制剂的海上油气田来说更有吸引力。墨西哥湾、北海、非洲西部等地区的气田已经开始采用这种化学抑制剂。但是,该方法也有局限性,主要是油相必须连续,因此只能用于含水率比较低的情况,最高含水率为 40% ~ 50%。这种局限性是由于流体中悬浮的大量固体碎屑改变了流体的流变学性质,也有可能与流体的流动型态有关。

如前所述,化学抑制剂的选择应该考虑设备限制和经济条件两方面。然而,现场生产条件也能限制化学抑制剂的选择。例如,Baker Petrolite 的研究项目结果表明,在恶劣条件下,可以使用反凝聚剂来防止水合物形成,与使用动力水合物抑制剂或活性抑制剂不同,反凝聚剂的用量不会随温度的降低而增加。因此,这种处理方法降低了抑制天然气水合物形成的成本。

抑制剂用量估算

抑制剂的最佳浓度是防止水合物形成的最小浓度。准确地预测这个最小浓度不仅是降低成本设计的要求,更有利于多相管流的运行。

注入系统的设计

化学抑制剂注入系统的设计是相当复杂的,包括抑制剂优选、确定必要的注入速率、泵的大小,以及管径的选择。对海上天然气集输系统,气田正常投产之前就要先选好抑制剂。因为影响抑制剂效果的诸多因素是未知的,包括盐水组分、温度和压力等,所以抑制剂的选择很困难。因此,目前注入系统的设计需要一个合理的多相流模拟程序来计算这些未知的必要参数。

Key to Exercises

Ⅰ.1. In general, two methods of preventing hydrate formation are applicable at the well site, namely thermal and chemical.

2. Heat conservation is accomplished through insulation. The design of such conservation systems typically seeks a balance among the high cost of the insulation, the intended operability of the system, and the acceptable risk level. The simplest is an external hot – water jacket, either for a pipe – in – pipe system or for a bundle. Other methods use either conductive or inductive heat tracing. There is concern over the reliability of conductive systems. An electrical resistance heating system may be desirable for long offset systems, where available insulation is insufficient, or for shut – in conditions. The ability to heat during production depends on the specific electrical heating implementation.

3. The inhibitor selection process often involves comparison of many fac-

tors, including capital/operating cost, physical properties, safety, corrosion inhibition, and gas dehydration capacity. However, a primary factor in the selection process is whether the spent chemical will be recovered, regenerated, and reinjected.

4. It was shown that under severe conditions, the required dosage of an antiagglomerator unlike thermodynamic and kinetic inhibitors does not increase as the degree of subcooling increases. Therefore, this method of treatment would be a cost – effective solution for the control of gas hydrates.

5. Proper design of an inhibitor injection system is a complex task that involves optimum inhibitor selection, determination of the necessary injection rates, pump sizing, and pipeline diameters. Inhibitors for a subsea gas transmission system are selected before gas production is started on the facility. This makes inhibitor selection difficult, as a large number of factors, including brine composition, temperature, and pressure, that affect the performance of inhibitors are unknown. Therefore, at this stage, an appropriate multiphase flow simulation package must be used to calculate some of the unknown necessary variables, which are required for injection systems design.

II. 1. A 2. B 3. B 4. D 5. A 6. C 7. C 8. D 9. A 10. B

III. 1. C 2. D 3. C 4. B 5. A 6. C 7. C 8. A 9. A 10. D

IV. 1. 保温系统的设计通常需要同时考虑材料成本(保温材料成本较高)、系统的可操作性、承受风险的能力三方面,在三者之间达到平衡。

2. 海底管线化学抑制剂注入系统的基础建设费用和运行费用都很高。

3. 对于海上气田,设法增加采出水中盐含量,也能降低水合物生成温度,然而仅仅靠这个方法是不够的。

4. 在恶劣条件下,可以使用反凝聚剂来防止水合物的形成,与使用动力水合物抑制剂或活性抑制剂不同,反凝聚剂的用量不会随温度的降低而增加。

5. 因此,目前注入系统的设计需要一个合理的多相流模拟程序来计算这些未知的必要参数。

V. 1. 化学抑制剂的选择通常需要对很多因素进行比较,包括基础建设和运行费用、天然气的物理性质、安全因素、防腐以及天然气的脱水能力。但是,首要考虑的因素是化学药剂能否回收、再生、循环注入。由于乙二醇这种抑制剂很昂贵,必须回收再生。无论是对于海上气田还是近海气田,乙二醇的回收再生装置都相当昂贵,而且需要占用很多空间。因此,需要研发一种新的、成本更低的、在低浓度下就能有效预防水合物形成的化学抑制剂。

2. 化学抑制剂注入系统的设计是相当复杂的,包括抑制剂优选、确定必要的注入速率、泵的大小,以及管径的选择。对海上天然气集输系统,气田正常投产之前就要先选好抑制剂。因为影响抑制剂效果的诸多因素是未知的,包括盐水组分、温度和压力等,所以抑制剂的选择很困难。

6.3 天然气脱水

导语

为确保天然气处理和输送安全,必须减少和控制天然气中的水含量。最常用的天然气脱水方法是液体干燥剂(甘醇)脱水法、固体干燥剂脱水法以及冷却法(气体冷却)。使用固体干燥剂有明显的经济和环境效益,降低了固定资产投资,减少了操作和维护成本,使挥发性有机化合物和有害气体污染降到最小。具体应用中,正确选择干燥剂是很复杂的事情。本文提供了理想干燥剂的性能以供参考。

课文

简介

天然气、伴生气和尾气中通常都含有以液相和气相形式存在的水,这些水主要来源于脱硫过程中的水溶液。工程经验表明,必须减少和控制气体中的水含量,以确保天然气处理和输送安全。进行天然气脱水的主要原因有以下几点:

(1)天然气在适合条件下能够与液态水或游离水结合形成水合物,阻塞阀门或管线;

(2)水能在管线里凝结,引起段塞流和腐蚀;

(3)水蒸气增大气体体积并降低气体热值;

(4)天然气销售合同和输气管线的技术规格中通常都要求天然气最大含水量为 $7lb/10^6ft^3$。

井口附近、集气管线及主要干线的重要位置处装有分液器,用来脱除井中气流带出的游离水。多级分离器也能够用来减少游离水。然而,脱除天然气中的水蒸气需要更复杂的处理方法。这些处理方法包括天然气"脱水",即通过降低天然气露点,使水蒸气从天然气中凝结出来。

天然气脱水有几种方法。最常用的方法是液体干燥剂(甘醇)脱水法、固体干燥剂脱水法以及冷却法(例如气体冷却)。前两种方法是将大量水分子转移到液体溶剂(甘醇)或结晶体(干燥剂)中。第三种方法利用冷却器使水分子凝结成液相,接着注入抑制剂防止水合物形成。脱水方法一般是在甘醇法和固体干燥剂法中选择。

甘醇脱水

在不同的气体干燥方法中,吸收法是最常用的一种方法。天然气中的水蒸气可以用液体溶剂来吸收。甘醇是使用最广泛的吸收溶液,其特性符合商业应用标准。一些甘醇溶剂已商业化应用,效果较好。

常用的甘醇及其使用说明如下:

(1)乙二醇(MEG):反应器中与天然气达到高蒸气压的平衡,易于进入气相,可以用作水合物抑制剂,因此在低于 50 ℉ 时能通过分离从天然气中回收。

(2)二甘醇(DEG):高蒸气压使其在反应器中大量损失,低分解温度所需要再凝温度较低(315～340 ℉),多数情况下不能得到足够纯的气体。

(3)三甘醇(TEG):最常使用,再凝结温度 340～400 ℉,纯度很高,反应器温度超过 120 ℉ 时,出现高蒸气损失,露点降低至 150 ℉ 可能发生汽提现象。

(4)四甘醇(TREG):比三甘醇昂贵,但在接触温度较高时损失较少,再凝结温度 400～430 ℉。

三甘醇是天然气脱水中最常使用的液体干燥剂,其特性满足大部分商业指标,具体如下:

(1)由于沸点和分解温度较高,三甘醇在常压汽提器中很容易再生成浓度为 98%～99% 的溶液;

(2)三甘醇初始理论分解温度在 404 ℉,而二甘醇只有 328 ℉;

(3)气化损失比乙二醇和二甘醇少,因此三甘醇在符合管线水露点指标要求情况下很容易再生成高浓度溶液;

(4)投资和生产成本较低。

固体干燥剂脱水

固体干燥剂脱水基于吸附原理。吸附是固体干燥剂界面和天然气中水蒸气附着的一种形式。由于吸附作用,水会在干燥剂表面形成一层极薄的液膜,这其中不发生化学反应。

固体干燥剂比甘醇干燥剂更有效,能把气体干燥到浓度小于 $0.05 \text{lb}/10^6 \text{ft}^3$。然而,为了减小固体干燥器的体积,经常先使用甘醇脱水装置去除大多数水分,可以把含水量减少到大约 60×10^{-6}(体积比)左右,这样会减小固体干燥剂的用量。

使用固体干燥剂替代甘醇干燥剂有明显的经济和环境效益,降低了固定资产投资,减少了操作和维护成本,使挥发性有机化合物和有害气体污染降到最小。

干燥剂性能

干燥剂性能常用单位质量干燥剂吸水的质量表示。干燥剂动态吸温性能取决于很多因素,如吸入气体的相对湿度、气体流量、吸附区温度、颗粒大小、干燥剂工作时间及污染程度、干燥剂自身。吸湿能力不随压力波动变化,除非压力会影响到前面提到的各种因素。下面是三个表示性能的术语:

(1)静态平衡吸水量:在平衡单元中没有流体流动时新干燥剂的吸水量(根据吸附等温线计算得出)。

(2)动态平衡吸水量:流体以工业流量流过干燥剂时干燥剂的吸水量。

(3)有效吸水量:是指考虑了吸水能力会随时间降低且所有干燥剂都不能充分利用情况的设计吸水量。

干燥剂的选择

市面上有各种各样的干燥剂可供选择,一些只对干燥气体有用,而另一些既能脱水也能去除重烃。具体应用中,正确选择干燥剂是很复杂的事情。在气体脱水过程中使用的固体干燥剂需要具备以下性质:

(1)平衡时吸收能力高。这可以减少所需吸附剂的体积,能够使用较小容器,减少设备投资和再生所需的热量。

(2)选择性强。这可以减少不希望除去的有价值组分的损耗,而且可以全面降低运行成本。

(3)再生容易。再生温度相对较低可减少能量需求和运行成本。

(4)压降低。

(5)机械性能良好(如高破裂强度、低磨损、低粉尘、稳定抗老化)。这些因素可降低吸收剂更换频率,减少生产中的停工损失,降低整体维护工作量。

(6)便宜,耐腐蚀,无毒,化学性质不活泼,体积密度高,吸水脱水体积变化不显著。

Key to Exercises

Ⅰ.1. The commonly used glycols are:(1) Monoethylene glycol (MEG);(2) Diethylene glycol (DEG);(3) Triethylene glycol (TEG);(4)Tetraethylene glycol (TREG).

2. TEG is by far the most common liquid desiccant used in natural gas dehydration, because it exhibits most of the desirable criteria of commercial suitability as following.

(1) TEG is regenerated more easily to a concentration of 98 – 99% in an atmospheric stripper because of its high boiling point and decomposition temperature.

(2) TEG has an initial theoretical decomposition temperature of 404 ℉, whereas that of diethylene glycol is only 328 ℉.

(3) Vaporization losses are lower than monoethylene glycol or diethylene glycol. Therefore, the TEG can be regenerated easily to the high concentrations needed to meet pipeline water dew point specifications.

(4) Capital and operating costs are lower.

3. Using desiccant dehydrators as alternatives to glycol dehydrators can yield significant economic and environmental benefits, including reduced capital cost, reduced operation and maintenance cost, and minimal VOC and hazardous air pollutants (BTEX).

4. The dynamic moisture sorption capacity of a desiccant will depend on a number of factors, such as the relative humidity of the inlet gas, the gas flow rate, the temperature of the adsorption zone, the mesh size of the granule, and the length of service and degree of contamination of the desiccant and not the least on the desiccant itself.

5. For solid desiccants used in gas dehydration, the following properties are desirable.

(1) High adsorption capacity at equilibrium.

(2) High selectivity.

(3) Easy regeneration.

(4) Low pressure drop.

(5) Good mechanical properties (such as high crush strength, low attrition, low dust formation, and high stability against aging).

(6) Inexpensive, noncorrosive, nontoxic, chemically inert, high bulk density and no significant volume changes upon adsorption and desorption of water.

Ⅱ. 1. D 2. C 3. A 4. B 5. C 6. A 7. D 8. B 9. D 10. C

Ⅲ. 1. A 2. A 3. D 4. C 5. B 6. C 7. B 8. A 9. D 10. C

Ⅳ.1. 天然气、伴生气和尾气中通常都含有以液相和气相形式存在的水,主要来源于脱硫过程中的水溶液。

2. 工程经验表明,必须减少和控制气体中的水含量,以确保天然气处理和输送安全。

3. 使用固体干燥剂替代甘醇干燥剂有明显的经济和环境效益,降低了固定资产投资,减少了操作和维护成本、使挥发性有机化合物和有害气体污染降到最小。

4. 一些只对干燥气体有用,而另一些既能脱水也能去除重烃。

5. 这些因素可降低吸收剂更换频率,减少生产中的停工损失,降低整体维护工作量。

Ⅴ.1. 井口附近、集气管线及主要干线的重要位置处装有分液器,用来脱除井中气流带出的游离水。多级分离器也能够用来减少游离水。然而,脱除天然气中的水蒸气需要更复杂的处理方法。这些处理方法包括天然气"脱水",即通过降低天然气露点,使水蒸气从天然气中凝结。

2. 天然气脱水有几种方法。最常用的方法是液体干燥剂(甘醇)脱水法、固体干燥剂脱水法以及冷却法(例如气体冷却)。前两种方法是将大量水分子转移到液体溶剂(甘醇)或结晶体(干燥剂)中。第三种方法利用冷却器使水分子凝结成液相,接着注入抑制剂防止水合物形成。脱水方法一般是在甘醇法和固体干燥剂法中选择。

6.4 结蜡处理

⬚ 导语

在油气集输系统中,有机固体物质(通常是蜡状物)的沉淀可能会中断生产,给多相油气集输带来严重的影响。沉淀物会导致地下或者地面设备堵塞和故障,特别是油气管线穿过北极地区或者其他温度较低的海域时。管线结蜡导致清蜡作业更加频繁,增加了风险。如果结蜡太厚,蜡块还会使管线容积减小并卡住刮蜡器。井筒和生产设备中结蜡会导致频繁关井或停产和其他操作问题。

⬚ 课文

结蜡

油气流结蜡过程可以看作热动力学的分子饱和现象。流体中溶解的石蜡分子最初处于一种混乱状态,然后石蜡分子在特定的热动力学状态下达到饱和而析出并沉淀下来。这个热动力学状态称为析蜡点或者凝点。这种现象和液体凝析现象中的露点是类似的,所不同的是结蜡过程是固体从液体中析出,凝析过程是液体从气体中析出。在结蜡过程中,由于冷却作用,树脂和沥青等大分子的动能降低到一定程度时,它们就会从溶液中析出并沉淀下来,而且分子结构不会破坏。如果给系统加热,增加分子的动能,这些分子就会在溶液中分散开来,重新形成稳定的悬浮液并做布朗运动。

抑制或预防结蜡

在原油流动管线中蜡分子和晶体混合物分散于并附着在井壁上形成蜡沉积。结蜡减少了管线的有效输送面积,增大了管壁粗糙度,导致压降增大。

结蜡的抑制或预防主要方法是注入专门的分子化合物,在高于浊点的温度下与石蜡分子相互作用,通过减少晶体与管壁间的吸附力影响结晶过程。由于原油流动时剪应力的作用,蜡分子晶体就会随油流带走。

当蜡晶体在管壁上形成时,上述机理就会发生。大多数充满液体的裸管(没有保温层)都属于这种情况。两个重要的前提条件是管壁快速传热和石蜡分子向管壁快速扩散。对于输送天然气/凝析油的管线,情况并非如此。虽然可以快速冷却,但是大部分气体在主流中流动冷却,因此,和主流液体形成石蜡晶体,由于重力的作用有沉淀和下降的趋势。而且,流体在裸管管线入口处未达到浊点的时候就已经有液体产生了,这样的液体黏度很高,与主流相比几乎不流动。在这种情况下,堵塞对管线清除蜡泥是有益的。

此外,这些流体基本上都吸收了用来防止水合物形成的甲醇。正好大多数(不是所有)化学剂具有改变蜡晶体的特性,它们与酒精和乙二醇是不相溶的。因此,在流动管线中抑制蜡晶体的堆积反而会增加蜡的积累。这种积累的蜡称为"蜡泥"。蜡泥具有黏性但可以移动,在足够的剪切力作用下可以流动。

现在对处理结蜡的两种化学方法进行简短的说明比较合适。有三种类型的蜡晶体:片状晶体、针状晶体、无定形晶体。

片状晶体或针状晶体由石蜡油形成,沥青质油主要形成无定形晶体。沥青质充当晶核使蜡晶体生成无定形晶体。片状晶体在显微镜下呈片状,针状晶体像针,无定形晶体没有固定形状并且通常看上去像小圆球。晶体与管壁的相互作用使晶体由无定形变成针状然后到片状。因此,使新生成的蜡晶体又小又圆,就像无定形晶体一样就可以了。

烃沉降形成的蜡晶体的特性,如浊点和倾点,受下面三方面因素的影响:

（1）晶体大小的改变：晶体从较大尺寸变为较小尺寸。

（2）成核作用抑制：抑制晶体的增长速度和最终尺寸。

（3）晶体类型或结构改变：晶体从一种类型变为另一种类型,例如,一个晶体从针状变为无定形。

蜡晶改性剂主要用于改变晶体的大小。烃中沉淀的正烷烃片状晶体加入蜡晶改性剂后受到抑制,在显微镜下看比以前更小了。晶体小,相对分子质量就小,在油中的可溶性就高。此外,分子小,互相碰撞或与管壁碰撞的能量也小。蜡晶改性剂通过把自身插入晶体来干扰正烷烃,并阻碍其生长。

需注意的是,蜡晶改性剂的另一个名字是降凝剂。顾名思义,蜡晶改性剂能够抑制蜡晶体的增长,从而有效降低原油的倾点,因此降低它们之间相互作用的强度。

蜡分散剂能够抑制蜡成核,并将蜡晶体的类型从片状或针状变为无定形。在抑制剂的作用下,这些蜡晶体比较小,而且更容易被流体带走。沥青质和胶质的存在有助于加强分散剂的作用。分散剂与沥青质和胶质相互作用并黏结,因此除去蜡晶体增长所需的晶核。当分散剂加到油中,温度高于浊点时,有时还能和沥青质及胶质相互作用并粘住它们,使其不能最先沉积下来,从而抑制浊点。分散剂通常倾向于把蜡颗粒分散到油水界面。

蜡晶改性剂和蜡分散剂都是很有用的化学剂,能够减少蜡的生成和沉淀,不过机理不同。蜡分散剂的分子质量和大小通常比蜡晶改性剂小得多,因此,根据其黏度和流动性特点,蜡分散剂通常更适合在低温下应用。此外,一些蜡分散剂可溶于甲醇和乙二醇,因此可同时注入。应根据所生产的烃类和设备选择清蜡剂。

结蜡修复

选择刮蜡之前,必须仔细考虑系统的性能,了解清管器工作的动力学特征,并懂得如何不断清除刮掉的蜡。这项工作难度很大,控制流体绕过清管器很难。因此,很多清管器作业最后都因为清管器卡住而失败。

刮蜡的分析和决策必须由真正的专家来做。从系统开始运行时就采用清管器作业成功的概率就高。尽管这样,还必须认真地监测系统的性能指数。建议首先考虑临时关井、化学剂浸泡,然后快速启动,这是最安全的选择。

Key to Exercises

Ⅰ. 1. The wax crystals reduce the effective cross – sectional area of the pipe and increase the pipeline roughness, which results in an increase in pressure drop. The deposits also cause subsurface and surface equipment plugging and malfunction, especially when oil mixtures are transported across Arctic regions or through cold oceans. Wax deposition leads to more frequent and risky pigging requirements in pipelines. If the wax deposits get too thick, they often reduce the capacity of the pipeline and cause the pigs to get stuck. Wax deposition in well tubings and process equipment may lead to more frequent shutdowns and operational problems.

2. Prevention or inhibition of wax deposition is mainly accomplished by injection of a special class of molecules that interact with paraffin molecules at temperatures above the cloud point and influence their crystallization process in a way that diminishes the attraction of the formed crystals toward the wall. The inhibited formed wax crystals are removed from the system by the shear forces caused by the flowing oil.

3. They are as follows. (1) Crystal size modification: modification of the crystal from larger sizes to smaller sizes. (2) Nucleation inhibition: inhibition of the growth rate of the crystal and its ultimate size. (3) Crystal type or structure modification: modification of the crystal from one type to the other.

4. A wax crystal modifier works primarily to modify the crystal size. Smaller crystals have lower molecular weights and thus higher solubility in oil. Furthermore, smaller particles have smaller energy of interaction among themselves and the pipe wall. It is noted that another name for a wax crystal modifier is pour point depressant (PPD). As the name implies, wax crystal modifiers are very effective at suppressing the pour points of crude oils because they suppress the wax crystal growth, thus minimizing the strength of their interactions.

5. We should consider carefully the system's performance to understand the dynamics of the moving pig and continuous removal of the wax cuttings ahead of it.

II. 1. B 2. A 3. C 4. B 5. A 6. D 7. A 8. C 9. B 10. D

III. 1. B 2. D 3. D 4. B 5. C 6. C 7. A 8. B 9. D 10. C

IV. 1. 在油气集输系统中,有机固体物质(通常是蜡状物)的沉淀可能会中断生产,给多相油气集输带来严重的影响。

2. 结蜡的抑制或预防主要方法是注入专门的分子化合物,在高于浊点的温度下与石蜡分子相互作用,通过减少晶体与管壁间的吸附力影响结晶过程。

3. 现在对处理结蜡的两种化学方法进行简短的说明比较合适。

4. 烃沉降形成的蜡晶体的特性,如浊点和倾点,受下面三个因素的影响。

5. 选择刮蜡之前,必须仔细考虑系统的性能,了解清管器工作的动力学特征,并懂得如何不断清除刮掉的蜡。

Ⅴ. 1. 油气流结蜡过程可以看作热动力学的分子饱和现象。流体中溶解的石蜡分子最初处于一种混乱状态,然后石蜡分子在特定的热动力学状态下达到饱和而析出并沉淀下来。这个热动力学状态称为析蜡点或者凝点。这种现象和液体凝析现象中的露点是类似的,所不同的是结蜡过程是固体从液体中析出,凝析过程是液体从气体中析出。

2. 蜡晶改性剂和蜡分散剂都是很有用的化学剂,能够减少蜡的生成和沉淀,不过机理不同。蜡分散剂的分子质量和大小通常比蜡晶改性剂小得多,因此,根据其黏度和流动性特点,蜡分散剂通常更适合在低温下应用。此外,一些蜡分散剂可溶于甲醇和乙二醇,因此可同时注入。

6.5 天然气压缩

🔲 导语

将天然气压缩可以提高管道运输压力,增加给定尺寸管道的天然气输送量,减少因摩擦引起的输送损失,而且,在长距离运输中可以不设增压站。常用的两种基本类型的天然气压缩机是往复式压缩机和离心式压缩机。往复式压缩机通常是用电动机或燃气发动机作动力,而离心式压缩机则是用燃气轮机或电动机来驱动。两种压缩机都各有优势,具体选择哪一种可基于本文提出的一些原则。

🔲 课文

天然气工业的方方面面都要用到压缩,包括气举、气体回注(为保持地层压力)、天然气集输、天然气处理(天然气在流程或系统中的循环)、气体输送及分配系统,以及罐装运输或储存。最近几年,提高管线输送压力是一个趋势。在一定的管线规格下,若输送压力较高,能增加天然气的运量,减少因摩擦引起的输送损失,而且,在长距离天然气输送中可以不设增压站。天然气输送中,常用的压缩机有两种基本类型:往复式压缩机和离心式压缩机。往复式压缩机通常是用电动机或燃气发动机作动力,而离心式压缩机则是用燃气轮机或电动机来驱动。

选择设备的关键参数有设备的寿命周期成本、资本成本、维修费用(包括检修和备用件)、燃料或能耗成本。设备的使用情况以及需求变化也是需要考虑的重要因素。轮机都能用管输天然气作燃料,电动机则必须有电力支持。由于参数较多,选择最佳驱动装置的工作十分复杂,在最后选定之前,要比较不同类型的驱动方式。经济可行性研究对确定一个方案的经济性并作出最佳选择是十分重要的。此外,必须确定压缩作业是否应多级压缩,是串联还是并联。

往复式压缩机

往复式压缩机中气体的压缩和排出是通过气缸中活塞的线性移动完成的。往复式压缩机用自动弹簧阀门,当阀门两侧存在一定压差时,阀门打开。进气阀门打开,气体流入气缸开始压缩。往复式压缩机容易使进出其中的流体产生明显的压力波动。因此,压缩机的上游和下游都必须安装压力缓冲器以避免破坏其他设备。在建站设计时,必须考虑压力缓冲器的压力损失(静态流压部分)。

由于排量和排出压力范围较大,往复式压缩机广泛应用于天然气处理行业。往复式压缩分高速往复式压缩机和低速往复式压缩机。一般高速往复式压缩机转速在 900~1200r/min,低速往复式压缩机转速在 200~600r/min。高速往复式压缩机一般是"可分离的",由压缩机身和驱动装置组成,中间由联轴器或变速箱连接。完整的装置是:动力气缸和压缩机气缸装在一个机架上,活塞安装在压缩机气缸上。低速往复式压缩机统筹设计为一个整体。

离心式压缩机

离心式压缩机由推进器及其后面的扩散器和回流腔(可能有)组成。压缩机体可以有一级或多级(能到 8 或 10)压缩。整个压缩机组可由一个或多个压缩机体组成,有时也包括一个变速箱。管道压缩机一般是单组,有一级或两级压缩。

离心式压缩机的工作原理与往复式压缩机不同,工作特点也不同。离心式压缩机广泛应用于化工厂、炼油厂、陆上和海上气举、注气、集气与天然气输送。离心式压缩机出口压力可高达 10000psi,覆盖了往复式压缩机流量/压力范围。离心式压缩机通常用涡轮机或电动机驱动。输气中应用的离心式压缩机的通常转速是:5000hp 的装置 14000r/min,2000hp 的装置 8000r/min。

压缩机对比

下面是往复式压缩机和离心式压缩机的区别。

往复式压缩机相对于离心式压缩机的优点有:

(1)适合体积流量较低和压缩比较高的情况;

(2)高压缩比条件下效率较高;

(3)小型设备(不超过3000hp)投资成本相对较低;

(4)对流体组分和浓度的敏感性较低。

离心式压缩机相对于往复式压缩机的优点有:

(1)在体积流量大、压力低的情况下较理想;

(2)只有一个运动部件,结构简单;

(3)在正常的工作范围内效率高;

(4)维护费用低,利用率高;

(5)单位工作面积下气体容量更大;

(6)不会产生震动和脉动。

压缩机选择

选择压缩机的基本原则包括以下几个方面:

(1)在不同工作条件下保持较高的工作效率;

(2)最大的配置灵活度;

(3)低维护成本;

(4)低运行费用;

(5)可接受的设备投资;

(6)利用率高。

然而,压缩机的其他要求和特点要根据每个项目及管道操作者经验的不同而定。实际上,压缩机选择主要是根据购买者提出的操作参数对机器进行设计。

选择的工艺设计参数如下:

(1)流量。

(2)气体组分。

(3)入口压力和温度。

(4)出口压力。

(5)组件安装:

① 对于离心式压缩机,包括串联、并联、多级主体、多级部件、中间冷却等;

② 对于往复式压缩机,包括气缸数量、冷却、流量控制策略。

(6)零部件数量。

在很多情况下,操作者根据其策略、排放要求、寿命周期成本预测等因素,就决定了是用往复式压缩机还是离心式压缩机,以及选用什么类型的驱动装置。但在选择压缩机时,还应对压缩机进行水力分析以保证最佳应用。实际上,可以为一个最可能和最常用的工况点选择压缩机。基于单一工况点选择压缩机时需仔细评定,确保提供充分的速度安全系数及波动安全系数,避免其他潜在严重情况的发生。选型确定后,按选型绘制压缩机特性曲线图以评价压缩机在其他工况条件下的性能。在许多情况下,通过管道水动力分析和油气藏研究,可以得到多个工况点。这些工况有常遇到的,也有特殊情况下发生的。掌握了这些知识,就可能根据期望目标使压缩机选择最优化,如实现最低的燃料消耗。

Key to Exercises

Ⅰ. 1. In gas transmission, two basic types of compressors are used: reciprocating and centrifugal compressors. Reciprocating compressors are usually driven by either electric motors or gas engines, whereas centrifugal compressors use gas turbines or electric motors as drivers.

2. The key variables for equipment selections are life cycle cost, capital cost, maintenance costs, including overhaul and spare parts, fuel, or energy costs. The units level of utilization, as well as demand fluctuations, plays an important role. While both gas engines and gas turbines can use pipeline gas as a fuel, an electric motor has to rely on the availability of electric power.

3. A reciprocating compressor is a positive displacement machine in which the compressing and displacing element is a piston moving linearly within a cylinder. Reciprocating compressors are widely utilized in the gas processing industries because they are flexible in throughput and discharge pressure range. However, a centrifugal compressor stage is defined as one impeller, with the subsequent diffuser and (if applicable) return channel. Centrifugal compressors are widely used in chemical plants, refineries, onshore and offshore gas lift and gas injection applications, gas gathering, and in the transmission of natural gas.

4. The basic principles for choosing a compressor consist of the following: good efficiency over a wide range of operating conditions, maximum flexibility of configuration, low maintenance cost, low life cycle cost, acceptable capital cost and high availability.

5. Other factors include the requirements and features of each project and specific experiences of the pipeline operator. The purchaser will define the oper-

ating parameters for which the machine will be designed. In many cases, the decision whether to use a reciprocating compressor or a centrifugal compressor, as well as the type of driver, will already have been made based on operator strategy, emissions requirements, general life cycle cost assumptions, and so on.

Ⅱ. 1. A 2. A 3. D 4. B 5. C 6. B 7. D 8. C 9. C 10. A

Ⅲ. 1. C 2. A 3. B 4. C 5. D 6. A 7. D 8. B 9. D 10. C

Ⅳ. 1. 选择设备的关键参数有设备的寿命周期成本、资本成本、维修费用(包括检修和备用件)、燃料或能耗成本。

2. 经济可行性研究对确定一个方案的经济性并作出最佳选择十分重要。

3. 完整的装置是:动力气缸和压缩机气缸装在一个机架上,活塞安装在压缩机气缸上。

4. 离心式压缩机的工作原理与往复式压缩机不同,工作特点也不同。

5. 然而,压缩机的其他要求和特点要根据每个项目及管道操作者经验的不同而定。

Ⅴ. 1. 天然气工业的方方面面都要用到压缩,包括气举、气体回注、天然气集输、天然气处理、气体输送及分配系统,以及罐装运输或储存。最近几年,提高管线输送压力是一个趋势。在一定的管线规格下,若输送压力较高,能增加输送天然气的量减少因摩擦引起的输送损失,而且,在长距离天然气输送中可以不设增压站。

2. 在很多情况下,操作者根据其策略、排放要求、寿命周期成本预测等因素,就决定了是用往复式压缩机还是离心式压缩机,以及选用什么类型的驱动装置。但在选择压缩机时,还应对压缩机进行水力分析以保证最佳应用。实际上,可以为一个最可能和最常用的工况点选择压缩机。基于单一工况点选择压缩机时需仔细评定,确保提供充分的速度安全系数及波动安全系数,避免其他潜在严重情况的发生。

References

［1］Ken Anold, Maurice Stewart. Surface Production Operations. 2nded. London：Butterworth – Heinemann，1999.

［2］SaeidMokhatab, William APoe, James GSpeight. Handbook of Natural Gas Transmission and Processing. New York：Elsevier,2006.

［3］Francis SManning, Richard EThompson. Oilfield ProcessingVolume two：Crude Oil. Tulsa：PennWell Books，1995.

［4］Francis SManning, Richard EThompson. Oilfield Processing Volume one：Natural Gas. Tulsa：PennWell Books，1991.

［5］George VChilingarian, John ORobertson, Jr Sanjay Kumar. Surface Operations in Petroleum Production IDevelopments In Petroleum Science 19A. New York：Elsevier，1987.

［6］George VChilingarian, John ORobertson, Jr Sanjay Kumar. Surface Operations in Petroleum Production IIDevelopments In Petroleum Science 19B. New York：Elsevier，1987.

［7］靳明三. 天然气集输与处理技术. 北京：石油工业出版社,2009.

［8］朱利凯. 天然气处理与加工. 北京：石油工业出版社,1997.

［9］塞德·莫克哈塔布，威廉 A·波，詹姆斯 G·斯佩特. 天然气集输与处理手册. 北京：石油工业出版社,2009.

［10］肯·阿诺德，毛瑞斯·斯图尔特. 油田地面工程:采出液处理工艺与设备设计. 马自俊，等译. 北京：中国石化出版社,2010.

［11］中国油气田开发志总编纂委员会. 中国油气田开发志. 北京：石油工业出版社,2011.

［12］白晓东,孙铁民,李冰,等.论油气地面工程十二五科技发展方向.石油规划设计,2012,23(2):1 – 4.

［13］白晓东,汤林,班兴安,等. 油气田地面工程面临的形势及攻关方向. 油气田地面工程，2012,31(10):9 – 10.

［14］孟昭宇，汪是洋. 橡胶密封件在油气地面工程的应用. 中国城市经济，2011(4):102 – 104.

［15］王军. 基于地面三维激光扫描技术的油气地面工程改扩建测量. 甘肃科技，2012,28(22):40 – 41.

［16］王全英. 油气地面集输工艺流程仿真系统的设计与实现. 天津：天津财经大学,2011.

［17］贾稳鹏,杨晓明. 油气田地面工程施工阶段安全质量管理标准化. 企业导报,2013(4):58 – 59.

［18］王岳昌. 探讨油气田地面工程质量管理. 中国石油和化工标准与质量,2012(9).

［19］贾稳鹏,杨晓明. 油气田地面工程施工阶段安全质量管理标准化. 企业导报,2013(4).

［20］岳春明,刘明. 油气田地面工程建设质量控制研究. 中国石油和化工标准与质量,2012(11).

［21］李秋忙,李庆,云庆,等. 油气田地面工程标准化设计引领技术研究. 北京：中国石油规划总院,2011.